T0135456

Study of an alternative phase field model for low interfacial energy in elastic solids

Anke Böttcher

Logos Verlag Berlin

λογος

Bibliografische Information der Deutschen Nationalbibliothek

Die Deutsche Nationalbibliothek verzeichnet diese Publikation in der Deutschen Nationalbibliografie; detaillierte bibliografische Daten sind im Internet über http://dnb.d-nb.de abrufbar.

Dissertation, Darmstadt 2021

ISBN 978-3-8325-5337-1

Logos Verlag Berlin GmbH
Georg-Knorr-Str. 4, Geb. 10, 2681 Berlin

Tel.: +49 (0)30 / 42 85 10 90
Fax: +49 (0)30 / 42 85 10 92

http://www.logos-verlag.de

Es erscheint immer unmöglich, bis es vollbracht ist.
(Nelson Mandela)

für meine geliebten Kinder

Smilla, Paul, Bela, Bibiane.

Summary. In this work, we focus on questions related to solid state phase transitions studied with the Allen-Cahn model, in use for more than 50 years, and the hybrid model, first published in 2005. The models are verified by theoretical considerations of energy decay properties. We present a discretization in time and choose a finite element discretisation in space. The resulting system is implemented in order to investigate the models numerically. We discuss the obtained results for the phase field models without coupled constitutive equations and also for a more complex model applied to linear elasticity. We show differences between the models in terms of numerical efficiency, interfacial energy, and study the behaviour of the hybrid model on small scales, which differs from the Allen-Cahn model.

Zusammenfassung. In dieser Arbeit befassen wir uns mit Fragen im Zusammenhang mit Festkörper-Phasenübergängen, die mit dem Allen-Cahn-Modell, welches seit mehr als 50 Jahren verwendet wird, und dem Hybrid-Modell, welches erstmals in 2005 veröffentlicht wurde, untersucht werden. Die Modelle werden durch theoretische Betrachtungen hinsichtlich ihres Energieverhaltens verifiziert. Wir präsentieren eine Diskretisierung in der Zeit und wählen eine Finite-Elemente-Diskretisierung im Raum. Das resultierende System wird implementiert, um die Modelle numerisch zu untersuchen. Wir diskutieren die erhaltenen Ergebnisse für die Phasenfeldmodelle ohne gekoppelte konstitutive Gleichungen und auch für ein komplexeres Modell, angewendet auf lineare Elastizität. Wir diskutieren Unterschiede zwischen den Modellen in Bezug auf die numerische Effizienz und die Grenzflächenenergie und untersuchen das gegenüber dem Allen-Cahn-Model abweichende Verhalten des Hybrid-Modells auf kleinen Skalen.

Danksagung. Ich danke Prof. Harald Garcke von der Universität Regensburg zutiefst dafür, dass er sich in allerletzter Minute bereit erklärt hatte, das Erstgutachten zu übernehmen. Seine konstruktiven und achtsamen Rückmeldungen haben mir während des Schreibens in der finalen Phase sehr geholfen. Gleichermaßen danke ich Prof. Dieter Bothe, der ebenfalls sehr kurzfristig bereit war, ein weiteres Gutachten zu schreiben und dadurch die Einreichung der Arbeit an der TU Darmstadt ermöglicht hat. Prof. Ralf Müller von der Technischen Universität Kaiserslautern danke ich insbesondere dafür, dass er mich 2007 in seine Arbeitsgruppe aufgenommen und mich erstmalig mit Phasenfeldmodellen und ihrer numerischen Umsetzung bekannt gemacht hatte und dass er sich jetzt bereit erklärte, als zweiter Gutachter aufzutreten. Ich danke Prof. Hans-Dieter Alber, der mich 2011 in seine Gruppe integriert und mit mir sein Hybrid-Model implementiert hat. Er hat mich darin ermutigt, eine Arbeit über dieses Thema zu schreiben. Nicht vergessen möchte ich an dieser Stelle die unersetzliche Frau Cramer aus dem Dekanat Mathematik, die mir bei allen Formalitäten geholfen und mich zum Durchhalten ermutigt hat.

Durch Betreuungskomplikationen hat die Richtung dieser Arbeit mehrfach in ihren Zielsetzungen geschwankt und es war schwierig, neue Anforderungen mit dem Bestehenden in Einklang zu bringen. Dann fehlte der Austausch, um von mir identifizierte Probleme diskutieren und lösen zu können und so fühlte ich mich streckenweise sehr auf mich alleine gestellt. Kurz vor dem Niederlegen meines Promotionsvorhabens hat mich die Neugier, das Geheimnis des „nicht verschwindenden Kreises“ zu ergründen, in Kontakt mit Prof. Garcke gebracht, dem ich für den Austausch in dieser Frage sehr dankbar bin.

Private Notwendigkeiten haben die parallele Weiterführung dieser Arbeit immer wieder in Frage gestellt und ich hätte diese Jahre niemals überstanden, wenn ich nicht gute Freund:innen und Unterstützer:innen gehabt hätte. Dazu zähle ich an erster Stelle Dr. Mirjam Walloth, die mich permanent ermutigt, mir unendlich viele wertvolle Hinweise gegeben hat und nie müde wurde, die Inhalte dieser Arbeit mit mir zu diskutieren. Ihr genaues Korrekturlesen und unser mathematischer Austausch haben mir sehr geholfen.

Gerne erinnerte Weggenoss:innen waren meine Zimmerkolleg:innen Dr. Paul Felix Riechwald, Dr. Natascha Kraynyukova und Dr. David Wegmann aus der Analysis sowie Dr. Sofia Erickson aus der Numerik, mit denen ich mich über Themen innerhalb und außerhalb der Mathematik unterhalten konnte. Aus der Zeit bei Prof. Ralf Müller erinnere ich mich gerne an Dr. Regina Müller, geb. Schmidt, die nach so langer Zeit bereit war, das Martensit-Kapitel Korrektur zu lesen und an Jun.-Prof. Charlotte Kuhn. Ich danke Graeme Hague und Martin Richard für die Hinweise hinsichtlich der englischen Sprache.

Innerhalb meines privaten Umfeldes gilt mein großer Dank Brigitte Röhling, genannt „Oma Bitte", die mich bereits während meines Studiums durch die liebevolle Betreuung meiner Kinder unterstützte und Bernd Röhling, der immer alles für uns getan hat. Mein Dank gilt ebenso meiner Mutter Dr.-med. Petra Böttcher und meinem leider kürzlich verstorbenen Vater Prof. Dr.-med. Heinz-Dietrich Böttcher, sowie meiner Schwester Dr. Trixi Böttcher und meinem Bruder Martin Böttcher, der tapfer mit seinem ohne-Dr. weiterlebt und der allerbeste Bruder ist, sowie seiner Freundin Devi.

Für ihren seelischen und praktischen Beistand danke ich meinen Freund:innen Priska Jahnke, Christl und Björn Harres, Anika Hartmann, Sonja Frey und Marcel Andres, Maritta Barbehön, Tamara Berger, Anika Boss, Sandra Brinkmann, meiner Schulfreundin Anja Metze, sowie den Sekretärinnen der Analysis Anke Meyer-Dörnberg und Christiane Herdler, meinem ehemaligen Lieblingslehrer Christian Behrens und meiner Geigenlehrerin Miriam Teuber. Ich freue mich auf eine Zeit mit mehr Zeit mit ihnen.

Sehr verständnisvoll und staunend interessiert („warum macht sie das?") waren meine wunderbaren medico Kolleg:innen Johannes Reinhard, Jens von Bargen, Gudrun Kortas, Anita Starosta und Anke Prochnau, sowie meine Studis Fynn Held, Max Jansen, Helena Lolies und Summer Vornwald, die für mich eine Arbeitsatmosphäre schufen, in der die eigentlich unmögliche Fertigstellung dieser Arbeit gelang. Und einen lieben Dank für das an mich Denken an meine besonderen Freunde Arash Assadulahi und Sebghatullah Oqab.

Mein allergrößter Dank geht jedoch an euch, meine zauberhaften Kinder: Paul, Bela, Bibiane und Smilla, die ihr Generationen von Babysittern ertragen musstet und mich immer wieder durch eure einzigartigen Persönlichkeiten überrascht. Ich liebe euch und mir ist bewusst, dass ich euch sehr viel zugemutet habe und ich hoffe, dass ihr mir verzeiht. Von nun an werde ich für euch Schach, Fotoshop und die Namen aller Pokémon (auswendig) lernen. Und ich freue mich unglaublich auf unseren gemeinsamen Urlaub nach dieser für uns alle so anstrengenden Zeit. Mit dir, Paul, der du still und heimlich in der Woche vor meiner Verteidigung mit Helena zusammengezogen bist, möchte ich sehr gerne Wiesbaden erkunden und freue mich jedes Mal, wenn du „nach Hause" kommst. Danke auch an Boncuk, unsere von mir widerwillig liebgewonnene

Hündin, die mich immer wieder zu Pausen in die Natur gezerrt hat. Ich kann einen Hund zum Promovieren sehr empfehlen.

Zuallerletzt und zuallermeist bedanke ich mich bei meinem solidarischen und auf seine Art sehr liebevollen Ehemann Wolfgang Bier, der dieses nie enden wollende Projekt teils kopfschüttelnd, aber immer wohlwollend und unterstützend begleitete. Er hat mir insbesondere in den letzten Wochen die Ruhe verschafft, die ich für den Abschluss dieser Arbeit brauchte und immer zu mir gehalten und ich freue mich darauf, den Rest unseres gemeinsamen Lebens endlich ohne die lästige Diss, dafür aber mit viel Heavy Metal, ausgebauten Bussen und eines Tages vielleicht einigen Enkelkindern zu verbringen. Danke, Wolle!

CONTENTS

NOTATION

$A : B$	$:= a_{ij}b_{ij}$; inner tensor product
$a \cdot d$	$:= a_i b_i$; inner vector product
$I_{a,b}$	$:= [a, b]$
$\langle v, w \rangle$	$:= \int_\Omega vw\,dx$; L^2 scalar product
$\partial_\tau Y_h^n$	$= \frac{1}{\tau}\left(Y_h^n - Y_h^{n-1}\right)$
$(\hat{\cdot})$	unknown nodal term
$[\,]$	jump brackets
$\langle\rangle$	mean-value brackets
$(\cdot)^+, (\cdot)^-$	boundary values at the sharp interface
$B(,)$	continuous bilinear form
$J()$	continuous functional
$l(,)$	continuous linear form
$L(,)$	continuous linear form
$H^1(\Omega)$	$:= \{v \in L^2(\Omega) \mid Dv \in L^2(\Omega)\}$
$H_0^1(\Omega)$	$:= \{v \in H^1(\Omega) \mid v = 0 \text{ on } \partial\Omega\}$
$L^2(\Omega)$	$:= \left\{f : \Omega \to \mathbb{R} : \int_\Omega \lvert f(x)\rvert^2\,dx < \infty\right\}$
$\mathcal{S}^3$	space of symmetric 3×3 matrices
V_h	finite dimensional subspace of $H^1(\Omega)$
$V_{h,0}$	finite dimensional subspace of $H_0^1(\Omega)$
W_h	finite dimensional subspace of $H^1(\Omega)^n$
$W_{h,0}$	$:= \{v_{hu} \in W_h \mid v_{hu} = 0 \text{ in } \partial\Omega\}$
$W_{h,D}$	$:= \{v_{hu} \in W_h \mid v_{hu} = u_D \text{ in } \partial\Omega\}$

notation in [4]

Γ	sharp interface
δ	constant
ζ	local coordinate
η	local coordinate
ν	asymptotic parameter
ξ	local coordinate
σ	stress
$\phi(x,t)$	indicator function
$\phi_{\mu\lambda}(x,t)$	indicator function

$\psi_2(S)$	double well potential
$\tilde{\psi}_1(S)$	double well potential
$\psi_S(S,\varepsilon)$	sum of $W_S(S,\varepsilon)+\psi_2'(S)$
ω_1	constant
ζ	$:=\frac{\xi}{\nu^{1/2}}$
D	tensor of linear elasticity
d_1	constant
$\hat{E}(S,\varepsilon)$	energy
$f()$	(non)linear function
L_1	constant
$P_n()$	orthogonal projection
$S^{(\nu)}(x,t)$	asymptotic solution approach for S
$S^{(\mu)}(x,t)$	asymptotic solution approach for S
$S_i(\eta,\zeta,t),\ i=0,2$	function in asymptotic solution approach
$S_i^{(\mu)}(x,t), i=1,2$	function in asymptotic solution approach
$\hat{S}(x,t)$	function in asymptotic solution approach
$\tilde{S}_i(x,t),\ i=1,3$	function in asymptotic solution approach
si	constant
$T^{(\nu)}(x,t)$	asymptotic solution approach for T
$\hat{T}(S,\varepsilon)$	stress component of ersatz stress
$\mathcal{U}$	interface neighbourhood
$u^{(\nu)}(x,t)$	asymptotic solution approach for u
$u_i(\eta,\xi,t)$	unknown displacement of asymptotic solution
$\hat{u}(x,t)$	displacement
$v(x,t)$	part of interface velocity
w	constant stress component of ersatzstress
$W_{S,ers}(S)$	derivative of elastic energy with respect to S

notation in [78]

$\kappa_G,\ \kappa_G$	calibration constant
c	order parameter
$\mathbb{C}_A,\ \mathbb{C}_M$	constant tensors of linear elasticity
$f(c)$	asymetric double well potential
G	interface energy density
L	interface width parameter
M	mobility factor

further notation

$\Phi()$	function of fixpoint iteration
Ω	open domain
Ω^+, Ω^-	subdomains of Ω
$\bar{\Omega}$	domain $\in \mathbb{R}^n$
$\partial\Omega$	boundary of Ω
α	parameter of general model
α_{heat}	constant
β	parameter of general model
$\Gamma(t)$	center line of $\Gamma_{\mathcal{U}}(t)$
$\Gamma_{\mathcal{U}}(t)$	diffuse interface area
$\hat{\Gamma}(t)$	sharp interface
$\tilde{\gamma}_{ij}$	components of $\tilde{\varepsilon}(u)$
$\epsilon\nabla u^{\epsilon}$	viscosity term of Hamilton-Jacobi equation
$\varepsilon(u)$	Cauchy strain tensor
ε_{L^2}	error term in [3]
ε_z	constant
ε_ν	constant
$\bar{\varepsilon}$	constant eigenstrain
$\bar{\varepsilon}_{ij}$	components of $\bar{\varepsilon}$
$\tilde{\varepsilon}(S,u)$	phase dependend strain tensor
$\tilde{\varepsilon}_{ij}$	components of $\tilde{\varepsilon}(S,u)$
κ_Γ	curvature parameter
λ	parameter of the Allen-Cahn model
μ	parameter of the Allen-Cahn model
ν	parameter of the hybrid model
ξ	constant (fundamental theoreme of calculus)
ρ	constant
ρ_2	constant
τ	time step
τ_0	constant
$\psi(S)$	double well potential
$\psi'(S)$	derivative of $\psi(S)$ with respect to S
ω_1	constant
a	domain width in numerical example
a, b	constants defining a compact interval $[a, b]$
a, b	placeholder in binomial formulas
a, b	placeholder in Young's inequality

B, E, F	optimisation parameters in [3]
$B_r(v)$	open ball of radius r and center v
b	twodimensional time- and space independent force
$b(x)$	time independent force
b_x, b_y	components of time independent force
$\bar{b}$	constant
$\tilde{b}$	constant
$\mathbb{C}$	tensor of linear elasticity with constant entries
$\mathbb{C}(S)$	phase dependend tensor of linear elasticity
$C_{\delta,i}, i = 1, 6$	constant
C_ε	constant
$C_{\psi,i}, i = 0, 5$	constant
$\mathbb{C}_i, i = 1, 2$	constant tensors of linear elasticity
C_P	constant
$C_{W,i}, i = 0, 2$	constant
C'	constant
C''	constant
C'''	constant
$\tilde{C}$	constant
$\tilde{C}_{min}$	constant
$\tilde{C}_P$	constant
$\tilde{\tilde{C}}_P$	constant
$\hat{C}(S, \varepsilon)$	eshelby tensor
$c(r)$	$:= cr$ with $c = constant$ (hybrid model)
c_1	parameter of the Allen-Cahn model
$c_A(r)$	$:= c_A r$ with $c_A = constant$ (Allen-Cahn model)
$c_H(r)$	$:= c_H r$ with $c_H = constant$ (hybrid model)
$\tilde{c}(r)$	$:= \tilde{c}r$ with $\tilde{c} = constant$ (Allen-Cahn model)
$D(S_h^n, S_h^{n-1})$	difference quotient of $\psi(S_h)$
d	width of interface
$d_i,\ i = 1, 3$	constant
$\bar{d}$	constant
$\tilde{\bar{d}}$	constant
dt	time step
dx	mesh size
$E(S, \varepsilon)$	free energy density
$\bar{E}(S, \varepsilon)$	free energy
$\bar{E}(S_h, \varepsilon_h)$	semi discrete free energy

$\hat{E}(S, \varepsilon)$	free energy term in definition of the Eshelby tensor
$\bar{E}(S)_{AC}$	energy value (Allen-Cahn model)
$\bar{E}(S)_H$	energy value (hybrid model)
$\bar{E}_{el}(S, \varepsilon)$	free elastic energy
$\bar{E}_{el,b}(S, \varepsilon)$	extended free elastic energy
$\bar{E}_{el,uh}(S_h, \varepsilon_h)$	fully discretised free elastic energy
e_{AC}	error (Allen-Cahn model)
e_H	error (hybrid model)
$e_i, i = 1, 3$	constant
e^{μ}_{res}	error (Allen-Cahn model)
e^{ν}_{res}	error (hybrid model)
err	error
$F(\hat{S})$	vector valued function (in Lemma 8)
$F(S^n_h, S^{n-1}_h, u^n_h)$	difference quotient of $W_S(S_h)$
F_{el}	vector of elastic forces
$F_I(\hat{S})$	scalar valued function at global node I (in Lemma 8)
$\tilde{F}()$	Lipschitz continuous function (in Lemma 6)
$f_1()$	(non)linear function (Allen-Cahn model)
$f_2()$	(non)linear function (hybrid model)
$\bar{f}$	constant
$\tilde{\bar{f}}$	constant
$\hat{f}$	vector of nodal forces
$g(r)$	$:= gr$ with $g = constant$ (Allen-Cahn model)
$\tilde{H}()$	invertible continuous function (in Lemma 6)
$h(\|\nabla S\|)$	reciprocal of the mobility function of the general model
$\underline{h}, \bar{h}$	constant
I	global node index
I	identity tensor
i	local nodal index
K	global element stiffness matrix
$\bar{K}$	global element stiffness matrix (with elasticity)
$\tilde{K}()$	Lipschitz continuous function (in Lemma 6)
k	iteration index
M_1	global mass matrix
M_{2mod}	global mass matrix (general model)
M_{2AC}	global mass matrix (Allen-Cahn model)
M_{2H}	global mass matrix (hybrid model)
$N_I(x)$	shape function for S

$N_I^u(x)$	shape function for u
N_t	number of time steps
n	space dimension
n	time step index
n	normal interface unit vector
nx	number of nodes per direction
p	parameter (shrinking circle)
$p\left(a_i(\hat{S}_I)^i\right)$	polynomial function
q	place holder of function h in Chapter 5
r	radius
r	place holder for e.g., a sum of functions
r_{eA}	width relation of interfaces
r_{eB}	width relation of interfaces
r_{st}	time derivative of radius r at t_{st}
r_t	time derivative of radius r
$r_{tAC,ers}$	interface velocity with ersatz stress (Allen-Cahn model)
$r_{tH,ers}$	interface velocity with ersatz stress (hybrid model)
$S(t,x)$	order parameter
S_0	initial condition for $S(x,t)$ at $t=0$
$S_h(x)$	semi discrete order parameter $\in C^1([0,T],V_{h,0})$
$S_{h,0}$	intitial condition for $S_h(x)$ at $t=0$
S_h^n	semi discrete order parameter at time step n
$S_{h,t}(x)$	$:=\frac{S_h^n-S_h^{n-1}}{\tau}$
S_t	time derivative of S (Chapter 7-10)
$\partial_t S$	time derivative of S (Chapter 1-6)
$\tilde{S}_h^n$	another semi discrete order parameter at time step n
$\hat{S}$	vector of nodal unknown order parameters
$\hat{S}^n$	vector of nodal unknown order parameters at time step n
$\hat{S}_I$	scalar nodal unknown order parameter at global node I
$\hat{S}_i$	scalar nodal unknown order parameter at local node i
$\hat{S}_k$	iterated vector of nodal unknown order parameters
$\hat{S}_t$	vector of time derivatives of nodal unknown order parameters
s_1	normal interface velocity (Allen-Cahn model)
s_{1err}	error term of normal interface velocity (Allen-Cahn model)
$s_{1\mu}$	part of normal interface velocity (Allen-Cahn model)
s_2	normal interface velocity (hybrid model)
s_{2err}	error term of normal interface velocity (hybrid model)
$s_{2\nu}$	part of normal interface velocity (hybrid model)

s_μ	normal interface velocity (Allen-Cahn model)
s_ν	normal interface velocity (hybrid model)
s^{el}_{AC}	normal interface velocity (elastic Allen-Cahn model)
s^{el}_{H}	normal interface velocity (elastic hybrid model)
$s_{AC,ers}$	normal interface velocity (ersatz Allen-Cahn model)
$s_{err,AC}$	error term (Allen-Cahn model)
$s_{err,H}$	error term (hybrid model)
$s_{H,ers}$	normal interface velocity (ersatz hybrid model)
s_{h_G}	normal interface speed in Garcke model
$s_{Mart.}$	normal interface speed in martensite model
$\hat{s}$	driving force due to elastic terms
$T(S,u)$	Cauchy stress tensor
t	time
t_{st}	point in time of stagnation
tol	constant
$u(x,t)$	displacement
$u_1(x)$	horizontal displacement boundary value
$u_2(x)$	vertical displacement boundary value
$u_D(x)$	Dirichlet border displacements
$u_h(x)$	semi discretised displacement
u_h^n	u_h at time step n
$\hat{u}$	vector of nodal displacements
$\hat{u}^n$	vector of $\hat{u}$ at time n
$\hat{u}_I$	unknown global nodal displacement
$u_{heat}(x,t)$	unknown temperature
$v(x)$	continuous function (Lemma 8)
$v_{hS}(x)$	semi discrete test function (general model)
$v_{hu}(x)$	semi discrete test function (elasticity)
v_S	test function (general model)
v_{SI}	nodal test function (general model)
v_u	test function (elasticity)
v_{uI}	nodal test function (elasticity)
$W(S,\varepsilon)$	elastic energy density
$W_S(S,\varepsilon)$	derivative of $W(S,\varepsilon)$ with respect to S
width_{AC}	interface width (Allen-Cahn model)
width_{H}	interface width (hybrid model)

PART I: BACKGROUND

1 Introduction

Almost a hundred years ago, Lev Davidovich Landau introduced the concept of an order parameter, indicating the state of a specific phase by a certain scalar value, in the context of superconductivity and phase transitions. Since then, a large amount of literature has been published around different order parameter concepts, applying the Allen-Cahn model to phase field problems in different gaseous states partially coupled to different material laws and other physical conditions.

This thesis compares this classical Allen-Cahn model to the new stated hybrid model applied to solid phase field problems and later on coupled to linear elasticity. Both models differ in terms of their numerical behaviour and their suitability to specific problems as for example material inclusions with high or low interfacial energy. The development of numerical methods for the simulation of phase transformations is explained proving the thermodynamic consistency of the discrete models. To confirm the theoretical results in [4] for the hybrid model, we extend the few simple numerical examples shown by the authors. Encountering special differences, we finally examine the behaviour of both models numerically applied to small martensite nuclei and explain the peculiarities of the hybrid model.

Remark 1.1. To aid understanding, we summarise contents of the original literature [1]–[6] to explain the correlation between the author's arguments and our investigations and to enable the reader to work with the original publications. It might be useful to consult the original literature for a deeper understanding of the analytical considerations of the hybrid model at some point.

Before we start with the mathematical theory, we want to say something about the appearance and the classification of phases and introduce the order parameter, we will use throughout this work.

Examples of phase fields. Phase field models are used to describe various phenomena in technology and nature. The existence of two or more different phases, partially even in different gaseous states, is an apparent characteristic of a melting glacier, a binary alloy below a critical temperature, a malign tumour or an oil slick. In many situations, we are interested in the changes of the involved phases in order to, e.g. forecast the danger of avalanches.

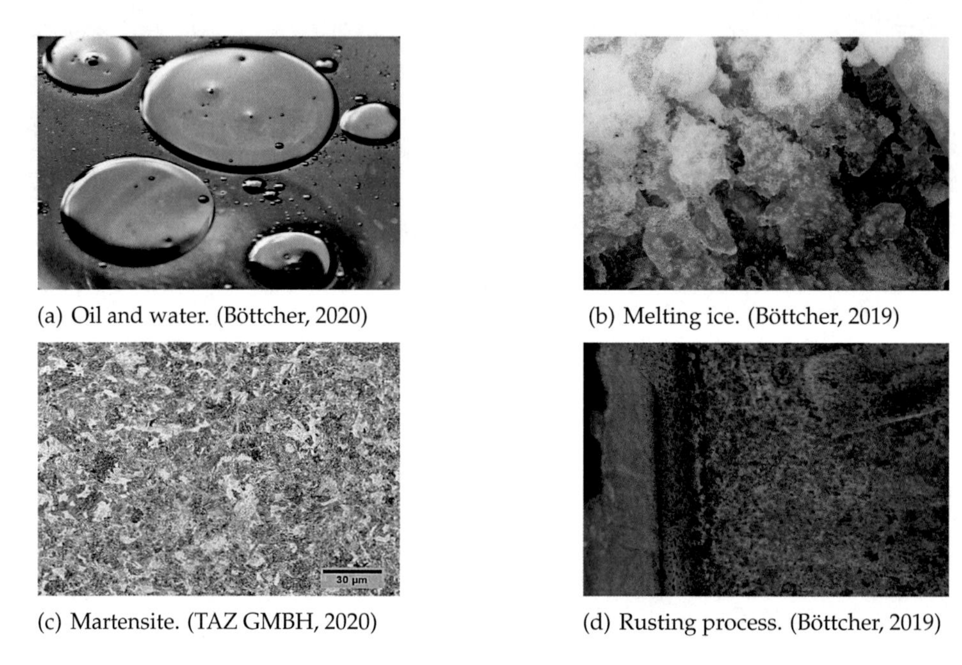

(a) Oil and water. (Böttcher, 2020)

(b) Melting ice. (Böttcher, 2019)

(c) Martensite. (TAZ GMBH, 2020)

(d) Rusting process. (Böttcher, 2019)

Figure 1.1: Examples of phases.

Phase transitions inducing topological changes at domain interfaces occur, e.g. in medical contexts in conjunction with the mutation of neoplasm. In materials science, diffusion of atoms or electrons cause interface changes with or without chemical reactions. The oxidation of iron with air and water leads to corrosion with phases of steel, rust, air and water. Snowflakes develop under a certain temperature from water vapour accumulating on condensation nuclei and in a martensite transformation the crystal lattice changes by sudden undercooling, building up new domains.

Phase field problems can be classified in different ways. Beneath the aggregate state combinations solid-solid, solid-liquid, solid-gaseous, liquid-liquid and liquid-gaseous there are other concepts of phase classification. The order parameter $S(x,t)$ can be a characteristic function for the concentration of a chemical component, the direction of a magnetic field, the polarisation in ferroelectric materials or the percentage of ionised particles in the earth's atmosphere. The kind of problem determines the methodical approach.

The topological development of a two-phase domain can be analysed by the spatial and temporal movement of the interface area. The movement of each material point belonging to this interface can be specified by its normal velocity. For such an analysis we need the more detailed definition of the above introduced order parameter.

Order parameter. The distribution of two phases can be modeled by an order parameter $S(x,t) : \mathbb{R}^n \times \mathbb{R} \to \mathbb{R}$, $x \in \Omega \subset \mathbb{R}^n$, $t \geq 0$ with dimension $n = 2$ or $n = 3$. We define the two subdomains $\Omega^+ = \{x : S(x,t) > \frac{1}{2}\} \subset \Omega$ and $\Omega^- = \{x : S(x,t) < \frac{1}{2}\} \subset \Omega$ corresponding to the location of the two phases. The pure phases are specified by a constant value of $S = 0$ and $S = 1$. They are separated by an interface region in which the order parameter increases continuously from 0 to 1 and we specify this region as the diffuse interface area

$$\Gamma_{\mathcal{U}}(t) = \{x : S(x,t) \in (0,1)\} \subset \Omega\,. \tag{1.1}$$

The set

$$\Gamma(t) = \{x : S(x,t) = 0.5\} \subset \Omega \tag{1.2}$$

denotes the centerline of the diffuse interface area $\Gamma_{\mathcal{U}}(t)$, see Fig. 1.2.

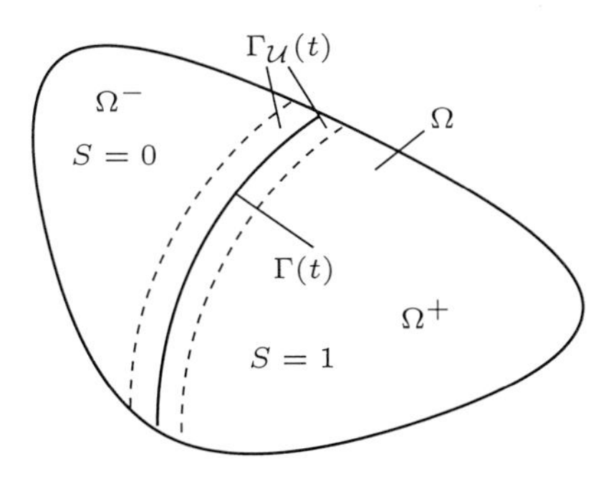

Figure 1.2: Two phases with smooth interface.

Another concept is the model of a sharp interface. We define it as a two-dimensional manifold $\hat{\Gamma}(t)$ embedded in the three-dimensional manifold Ω.

In two dimensions we can, in contrast to the diffuse interface $\Gamma_{\mathcal{U}}(t)$ with the centerline $\Gamma(t)$, imagine the sharp interface as a line on which the value of the order parameter jumps. The diffuse interface $\Gamma_{\mathcal{U}}(t)$ becomes very narrow when the asymptotic parameters, which we will explain in the following, approach zero. Then the so called limit model of the diffuse interface $\Gamma_{\mathcal{U}}(t)$ goes towards the sharp interface $\hat{\Gamma}(t)$, which means that $\Gamma(t)$ agrees with $\hat{\Gamma}(t)$.

Both kinds of phase transition are found in nature. Examples of a sharp interface are interfaces in martensite materials. The lattice is either in the metastable face-centered cubic austenite phase or in the stable body-centered cubic martensite phase. However, since the lattices of the martensitic and austenitic phases do not match, distortions occur, which lead to the formation of an interstitial lattice in a narrow area. Therefore, we can not speak of a real sharp interface, even though the interface is very narrow.

An example of diffuse interfaces are magnetic Bloch walls between magnetic domains. The direction of the magnetisation can not abruptly turn in the opposite direction, which is why magnetic transition zones form diffuse interfaces. Whether a physical interface is actually sharp or not also depends on the scale we choose to study the material (e.g. on the atomic level).

In numerical simulations, we often need to resolve the interface by defining nodes and, e.g. Finite Elements within. Thus, the width of the numerical diffuse interface region might (related to the respective domain) be wider than the interface of the research object.

In solid-solid phase field systems, on which we focus on in the present work, the evolution of $S(x,t)$ is characterised by a system of partial differential equations. It may depend on material properties such as Young's modulus, eigenstrain, crystal lattice constants and more. Boundary conditions like pressure, heat and other outer forces can be applied to the solid material in order to induce phase transformation processes. Furthermore, there are inner processes like chemical reactions or diffusion, which influence the distribution of the phases and thereby the topology and the properties of the material.

Historical embedding. Solid phase field problems can be described by phase field models found in literature already in the first half of the last century. Lev Landau presented a concept for phase transition in 1937, introducing the Landau potential and an order parameter, see [65], [63], [64], [25] and the references given therein. In 1950 Lev Landau and Witali Ginzburg used the concept of an order parameter describing the change from normal electrical conductivity to superconductivity with the Ginzburg-Landau-theory, see [48].

In 1979, John W. Cahn and Samuel M. Allen published their first work on the Allen-Cahn equation, see [7]. In this basic work, the time dependent change of an order parameter associated with the antiphase boundary motion is compared to the interface velocity depending on local principal curvatures. The related theory of mean curvature flow or mean curvature motion was discussed in many publications, see [29], [37], [36], [44], [57] and [32]. Since 1957 a huge amount of literature was published on Allen-Cahn type equations, e.g. [18], [70], [27], [58] and [76].

Solid-solid phase transformations are often caused by internal and external loads and then the phase field equations are coupled with constitutive material equations. These can represent the effects of lattice distortion, elastic and inelastic strains caused by heat, pressure and other loads, see [73], [72] and the references given therein. The Allen-Cahn equation was originally developed for iron-aluminum alloys and later on applied to other problems, inter alia martensite transformations, see [78], [77], crack extension, see [89], [80] and ferroelectric materials, see [16], [79], [83] and [85]. As mentioned before, Allen-Cahn type phase field equations can also deal with solid-fluid problems, e.g. the Stefan problem related to the melting and freezing of ice, see [20] and the references given therein.

The Allen-Cahn equation is a classical reaction-diffusion equation, see [61], [68] and a partial differential equation of (semilinear) parabolic type. It can be derived as a gradient flow of an associated free energy, see [14].

A good overview of publications on the Allen-Cahn equation, sometimes denoted as time-dependent Ginzburg-Landau equation, see [42], on many aspects such as asymptotic behaviour, error estimates, existence and uniqueness of solutions up to coupled problems is given in [84] and [26]. A summary of a general phase field theory and different limit models can be found in [24].

In 2005, the hybrid model was published in [6] and advanced in further works of the author's, see [1]–[5]. In [4], the hybrid model is explained as a model that replaces the (rescaled) mobility function of the Allen-Cahn model, regulating the propagation velocity of the interface by a term depending on the gradient of the order parameter.

In the following sections we will introduce the Allen-Cahn model and the hybrid model in particular. We will point out commonalities of and differences between both models also by means of numerical examples in Part II and Part III.

2 Elastic Allen-Cahn model

Fig. 1.2 shows a model for a system with two phases. Here, the order parameter has values $S \in [0,1]$ and both phases are separated by an interface given by Eq. (1.1). To describe the movement of the (diffuse) interface $\Gamma_{\mathcal{U}}(t)$, the later named Allen-Cahn model was 1979 introduced in [7] by Samuel M. Allen and John W. Cahn, based on earlier works [8].

The Allen-Cahn formulation in [7] had no coupling to constitutive laws representing inner and outer forces. Anticipating the purpose of the third part of the present work, we start with the formulation of an elastic phase field model, which includes an additional term due to elastic effects.

We start with the definition of a free energy that leads to the respective phase field equation. In [4], the free energy density function of the Allen-Cahn model, coupled to linear elasticity, is given as

$$E(S,\varepsilon) = W(S,\varepsilon) + \frac{1}{\mu^{1/2}}\psi(S) + \frac{\mu^{1/2}\lambda}{2}|\nabla S|^2 \tag{2.1}$$

with $S = S(x,t)$ and $\varepsilon = \varepsilon(u(x,t))$, defining the free energy

$$\bar{E}(S,\varepsilon) = \int_{\Omega} E(S,\varepsilon)\,dx\,. \tag{2.2}$$

The scalar-valued elastic energy density function

$$W(S,\varepsilon) = \frac{1}{2}(\varepsilon - \bar{\varepsilon}S) : (\mathbb{C}(\varepsilon - \bar{\varepsilon}S)) \tag{2.3}$$

depends on the symmetric linear Cauchy strain tensor

$$\varepsilon(u) = \frac{1}{2}\left(\nabla u(x,t) + (\nabla u(x,t))^{\mathsf{T}}\right) \in \mathcal{S}^3 \tag{2.4}$$

with $\mathcal{S}^3$, the space of symmetric 3x3-matrices. It is valid for small deformations and depends on the displacement $u(x,t) \in \mathbb{R}^3$ and we state

Assumption 1. $||\varepsilon(u)|| \leq C_\varepsilon, \quad C_\varepsilon \in \mathbb{R}_0^+.$

The tensor $\mathbb{C} : \mathcal{S}^3 \to \mathcal{S}^3$ of linear elasticity is a positive definite mapping and describes the resistance of a material against inner and outer forces like eigenstrains and pressure. We will later define a linear dependency on the order parameter to adjust $\mathbb{C}$ to the respective phase.

The scalar-valued double well potential with minima in 0 and 1 is in [4] defined by

$$\psi = 4S^2(1 - S)^2 \tag{2.5}$$

and $\lambda \in \mathbb{R}^+$ and $\mu \in \mathbb{R}^+$ are constant parameters.

Before we explain the meaning of these parameters in detail, we will show the correlation between the free energy function Eq. (2.2) and the Allen-Cahn model.

In the following, we use the notation

$$L^2(\Omega) := \left\{ f : \Omega \to \mathbb{R} : \int_\Omega |f(x)|^2 \, dx < \infty \right\}$$

for the vector space consisting of all Lebesgue-measurable functions, whose squares can be integrated and we define the Sobolev space

$$H^1(\Omega) := \{ v \in L^2(\Omega) \mid Dv \in L^2(\Omega) \}$$

with Dv denoting the weak derivative of v. Based on these definitions, we have the subspace

$$H_0^1(\Omega) := \{ v \in H^1(\Omega) \mid v = 0 \text{ on } \partial\Omega \}$$

for the later given homogeneous boundary conditions.

Remark 2.1. In the course of the paper we will sketch proofs in terms of existence and uniqueness and in Part III a general phase field model coupled with linear elasticity, representing the Allen-Cahn model and the hybrid model, will be implemented numerically. To keep both, the proofs and the implementation, simple, we assume that we always first calculate the displacements and strains of the elasticity equations and analyse and implement the phase field model with these known quantities. Therefore, the phase field formulations within the coupled elasticity do not contain any unknown displacement.

Lemma 1 (*Gradient flow*). *Let Eq. (2.2) be the free energy for the elastic phase field problem and assume u and S to be given and to be sufficiently smooth for the following correlations. The negative gradient flow of the free energy with respect to a weighted metric yields the elastic Allen-Cahn equation in terms of*

$$\partial_t S = -c_A \left(W_S(S,\varepsilon) + \frac{1}{\mu^{1/2}} \psi'(S) - \mu^{1/2} \lambda \Delta S \right) \tag{2.6}$$

with homogeneous Dirichlet or homogeneous Neumann boundary conditions for S.

Proof. The variation of the global formulation of the free energy Eq. (2.2) with known displacements in direction of $v \in H_0^1(\Omega)$ is defined by

$$\frac{\delta \bar{E}}{\delta S}(S)(v) = \frac{d}{d\rho} \bar{E}(S + \rho v)_{|\rho=0}\,, \quad \rho \in \mathbb{R}\,, \tag{2.7}$$

leading to

$$\frac{\delta \bar{E}}{\delta S}(S)(v) = \int_\Omega W_S(S,\varepsilon) v + \frac{1}{\mu^{1/2}} \psi'(S) v + \mu^{1/2} \lambda \nabla S \cdot \nabla v \, dx\,. \tag{2.8}$$

We denote by W_S the partial derivative of the elastic energy function $W(S,\varepsilon)$ with respect to the order parameter S and we define $\psi'(S)$ as the derivative of the double well energy function $\psi(S)$ with respect to S.

The Allen Cahn equation arises from the negative gradient flow of a related free energy with respect to a weighted metric, see [32]. Together with Eq. (2.8) we obtain

$$\langle \partial_t S, v \rangle = -c_A \int_\Omega W_S(S,\varepsilon) v + \frac{1}{\mu^{1/2}} \psi'(S) v + \mu^{1/2} \lambda \nabla S \cdot \nabla v \, dx \tag{2.9}$$

$$\forall v \in H_0^1(\Omega) \text{ and } t > 0\,,$$

see [44].

Integrating by parts, applying homogeneous boundary conditions and using the fundamental lemma of calculation of variations, see [31], the elastic Allen-Cahn equation (2.6) results and the correlation between Eq. (2.8) and Eq. (2.9) proves the Lemma.

□

Referring to the rescaling in [4], we have

$$c_A(r) = \frac{f_1}{(\mu\lambda)^{1/2}}(r)\,. \tag{2.10}$$

As explained in the original literature, the function f_1 is possibly non-linear, but the authors assume it to be linear by setting

$$f_1(r) := \tilde{c}r\,, \quad \tilde{c} \in \mathbb{R}^+\,, \tag{2.11}$$

yielding

$$c_A(r) = \frac{\tilde{c}}{(\mu\lambda)^{1/2}}r\,. \tag{2.12}$$

Inserting Eq. (2.10) into Eq. (2.6) leads to the more specific formulation

$$\partial_t S = -\frac{1}{(\mu\lambda)^{1/2}} f_1\left(W_S(S,\varepsilon) + \frac{1}{\mu^{1/2}}\psi'(S) - \mu^{1/2}\lambda\Delta S\right)\,. \tag{2.13}$$

We denote the factor in front of the brackets as mobility, scaling the velocity of the order parameter $\partial_t S$. A case of a non-linear mobility was discussed in [13] in the context of the Cahn-Hilliard equation. We will not regard a non-linear mobility in the present work.

Inserting Eq. (2.12) into Eq. (2.6), we formulate the elastic Allen-Cahn phase field equation with a linear mobility, we will regard in the following, as

$$\partial_t S = -\frac{\tilde{c}}{(\mu\lambda)^{1/2}}\left(W_S(S,\varepsilon) + \frac{1}{\mu^{1/2}}\psi'(S) - \mu^{1/2}\lambda\Delta S\right)\,. \tag{2.14}$$

In addition, initial- and boundary conditions are required to completely describe the evolution of the phase fields. We assume, following [4],

$$\begin{aligned} S(x,t) &= 0, \quad x \in \partial\Omega\,, \quad t \geq 0, \\ S(x,0) &= S_0, \quad x \in \Omega \end{aligned} \tag{2.15}$$

throughout this work.

Next, we will explain the meaning of the parameters λ and μ in Eq. (2.14). Therefore, we need an understanding of interfacial energy.

Interfacial energy in the context of phase boundaries. Without additional inner and outer forces, the particles inside a bounded domain are first balanced. Dominating surface tension causes shrinking interfacial areas in order to reduce the number of the unbalanced outer particles. Thus, nucleus inclusions in technical alloying applications round out and shrink. The formation of minimal surfaces can be observed in the example of soap bubbles, where the interfacial energy is related to the better known surface energy, see Fig. 2.1.

(a) soap bubbles t_0 (Böttcher, 2017)

(b) soap bubbles t_1 (Böttcher, 2017)

Figure 2.1: Minimisation of surface energy.

In contrast, additional inner and outer forces can counteract the interface minimising effects. The rearrangement or transformation of atomic structures can cause distortions of the well-balanced lattice and elastic forces gain influence. Their relation to the interfacial energies controls the corresponding phase evolutions. So, particles may grow and coarsen due to outer mechanical forces or inner misfittings, see [46], [45], [42].

In the present work, we understand λ as a interfacial energy parameter, depending on the respective material and the related effects such as, e.g. the dislocation of atoms close to the interface and their imbalance of forces, see Fig. 2.2. We will assume λ as a temperature independent constant.

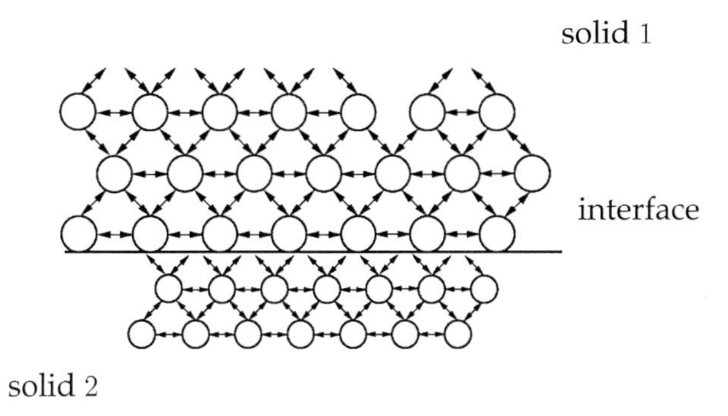

Figure 2.2: Interfacial energy.

The parameter μ is small and stands in the denominator in front of the derivative of the double well potential in Eq. (2.14) . Choosing a constant λ, a smaller μ increases the influence of the double well potential that separates the phases and thus, the interface width decreases. We will explain in Chapter 5.4 why a larger μ increases the numerical solution error for the normal interface velocity.

In general, there is no relation between the numerical and the physical width of a diffuse interface. In numerical simulations, the width of the diffuse interface depends on the parameter μ. This fact is important because the numerical resolution of the interface region needs about five grid points for sufficient accurate results, observed in numerical experiments, see [10], [75] and Chapter 7.

Interpretation of the energy functions. Omitting the term W_S in Eq. (2.14), the phase field behaviour and the movement of the interface $\Gamma_{\mathcal{U}}(t)$ depend on the curvature, the interface energy, controlled by the value of λ, and the double well potential ψ, scaled by the parameter μ. The derivative of the elastic energy W_S couples the equations of linear elasticity to the phase field model. Thus, inner as well as outer forces have an impact on the interface movement. Moreover, the phase field models can be coupled to other material laws, e.g. describing piezoelectricity, see Appendix 9.

The number and the values of the minima of the double well potential, here given by Eq. (2.5), depend on the given situation: the number and the influence of the involved phases. Static solutions minimise the free energy and therefore push the order parameter into one of these minima, promoting phase separations. For this reason, ψ is also called separation potential in literature. The (double well) potential has to be adjusted to the specific problem, see [78] and Chapter 10. For phase field systems with more than two phases multiple order parameter can be used. This situation is not discussed in this work.

Sharp interface limit: The limit model of the sharp interface normal velocity of the Allen-Cahn model

$$s = \frac{\tilde{c}}{c_1}\left(n \cdot [\hat{C}]n + \lambda^{1/2} c_1 \kappa_\Gamma\right), \quad \frac{\tilde{c}}{c_1}(r) = \frac{\tilde{c}}{c_1} r, \quad \frac{\tilde{c}}{c_1} \in \mathbb{R}^+ \tag{2.16}$$

with

$$c_1 = \int_0^1 \sqrt{2\psi(\Theta)}\, d\Theta \tag{2.17}$$

was defined within the asymptotic solution analysis in [4].

The brackets $[(\cdot)] = (\cdot)^+ - (\cdot)^-$ indicate the jump of the value of the respective term at the sharp interface. The indices $+$ and $-$ mark the respective sides.

In [4], approaches of asymptotic solutions were inserted into the belonging system consisting of the Allen-Cahn equation coupled to linear elasticity. The inner and the outer expansion, meaning the parts of the asymptotic approaches that are valid in the respective phase at both sides of the interface region, were matched within the diffuse interface region. Residual terms of the resulting system had to be estimated. Within these estimations, the expression (2.16) arised, describing the normal velocity of the diffuse interface $\Gamma_{\mathcal{U}}(t)$, converging to the normal velocity of the sharp interface $\hat{\Gamma}(t)$ by sending μ to zero.

The dependence of this approximation of the sharp interface velocity on the choice of μ was only clarified in a further publication [2]. Similar approaches can be found in [76], [43] and [34].

The formulation (2.16) contains the Eshelby tensor, which is in general given as

$$\hat{C}(S,\varepsilon) = \hat{E}(S,\varepsilon)I - \varepsilon(u)^{\mathsf{T}} T(\varepsilon(u)) \tag{2.18}$$

with the particular free energy

$$\hat{E}(S,\varepsilon) = W(S,\varepsilon) + \psi(S)\,, \tag{2.19}$$

the identity tensor $I \in \mathbb{R}^{3\times 3}$, the strain tensor ε, given by Eq. (2.4), and the definition of the Cauchy stress tensor as

$$T(\varepsilon(u)) = \mathbb{C}\varepsilon(u) \qquad \in \mathcal{S}^3 . \tag{2.20}$$

We will later define a dependency of T on the order parameter S.

The Eshelby tensor represents the energy-momentum tensor in the context of configurational forces, see [62], [74]. The term $[\hat{C}]$ stands for the jump of the Eshelby tensor, which is zero in the homogeneous phase and has non-zero-values at the sharp interface $\hat{\Gamma}(t)$, see [71], [1] and the references given therein. The normal interface unit vector $n(x,t)$ points into the direction of the phase indicated by $S = 1$. The constant c_1, given by Eq. (2.17), is a function of the double well potential ψ and κ_Γ is twice the mean curvature. Dropping the Eshelby-term in Eq. (2.16) yields a mean curvature motion depending on interface properties only, omitting inner and outer forces.

Now that we have explained all terms contained in the Allen-Cahn model, as well as the meaning of the associated energy functions and the (sharp) interface velocities, we conclude this chapter with some final remarks on the literature and the well-posedness of systems associated with the Allen-Cahn model.

The Allen-Cahn model is related to the Cahn-Hilliard equation describing phase separation under mass conservation [49], [17]. Here, instead of chemical reaction and diffusion, only diffusion influences the interface evolution. The modeling of mass conservation leads to a fourth-order partial differential equation.

The Cahn-Hilliard and the Allen-Cahn equation can be treated numerically in a similar way. One approach is to determine a coupled system of two second-order partial differential equations, replacing the second derivative of the order parameter by a new unknown, see [33]. In [46], [45], the Cahn-Hilliard model coupled to linear elasticity is given by an elliptic-parabolic system and the author shows existence and uniqueness of variational solutions under special assumptions. In [54], this proof is adapted to the elastic Allen-Cahn model by similar considerations.

Eq. (2.14) has no analytical solution in general. Therefore, numerical schemes have to be used to find solutions of the order parameter, describing the movement of the interfaces. There are many publications on numerical schemes applied to the Allen-Cahn problem with or without elastic energy, e.g. [66], [27]. An interesting approach, known as the splitting method, is proposed in [82].

A posteriori error estimates are found, among other references, in [60], [39] and [38]. Comparisons of different explicit and implicit implementation schemes connected to energy decay properties are given in [40], [69], [90].

A basic aspect of this work is that the choice of μ defines the width of the diffuse interface in numerical simulations. The numerical interface has to be resolved finely to obtain appropriate results. Therefore, the size of μ has a strong impact on the computational effort but also on the error of the simulation results. In Chapter 5.4 we will estimate the numerically developed width of the transition zone between the two phases based on a local scaling at the sharp interface and we will verify our predictions by numerical tests in Chapter 7.

3 Elastic hybrid model

In contrast to the Allen-Cahn model, the hybrid model was published only recently and so far there is almost no literature beyond the original publications on it. This work aims to fill this gap a little and will extend and supplement the investigations started in [1]–[6]. For this purpose, we will first introduce the model.

In 2005, Hans-Dieter Alber and Peicheng Zhu published their first work about the hybrid model [6], followed by further publications on analytical properties and asymptotic solution analysis, see [3]–[5].

To describe the behaviour of the order parameter S, the hybrid model was given by

$$\partial_t S = -f_2(W_S(S,\varepsilon) + \psi_2'(S) - \nu\Delta_x S)|\nabla_x S| \,. \tag{3.1}$$

The function $f_2(r)$ in Eq. (3.1) can be non-linear like the function $f_1(r)$ in Eq. (2.13). Here, we will only define it as a linear function.

In [4], the authors gave some simple examples of a two-dimensional numerical implementation at the end of their publication, comparing the hybrid model to the Allen-Cahn model. In the present work, we will extend their tests to some more complex numerical examples, assuming, e.g. $f_2(r) = cr$, with a constant $c \in \mathbb{R}^+$. We drop the index x and write Eq. (3.1) as

$$\partial_t S \;=\; -c\,(W_S(S,\varepsilon) + \psi_2'(S) - \nu\Delta S)|\nabla S| \,. \tag{3.2}$$

The Laplacian term, multiplied with the parameter $\nu \in \mathbb{R}^+$, determines the width of the diffuse interface, but in a less restrictive way as μ for the Allen-

Cahn model. This will be an important aspect of the simulations in Chapter 7 and will be explained in Chapter 5.4. For $\nu \to 0$ the solution converges to the sharp interface limit of the hybrid model.

Similar as explained in Chapter 2, the hybrid model can be derived as a gradient flow for a specific free energy with or without coupled constitutive equation terms. The associated free energy density function for the elastic hybrid model is given by

$$E(S,\varepsilon) = W(S,\varepsilon) + \psi_2(S) + \frac{\nu}{2}|\nabla S|^2 \tag{3.3}$$

with $S = S(x,t)$ and $\varepsilon = \varepsilon(u(x,t))$, leading to the free energy function in terms of

$$\bar{E}(S,\varepsilon) = \int_\Omega E(S,\varepsilon)\,dx\,. \tag{3.4}$$

The proof is performed like in Chapter 2.

Lemma 2 (*Gradient flow*). *Let the free energy for the elastic phase field problem be given by Eq. (3.4) and assume u to be known and S to be sufficiently smooth. Then the gradient flow of the free energy leads to the elastic hybrid model equation*

$$\partial_t S = -c\,(W_S(S,\varepsilon) + \psi_2'(S) - \nu\Delta S)|\nabla S|\,, \quad c(r) = cr, \quad c \in \mathbb{R}^+ \tag{3.5}$$

with homogeneous Dirichlet or homogeneous Neumann boundary conditions for S.

Proof. The variation of the free energy Eq. (3.4) goes along the lines of the proof of Lemma 1 leading to the hybrid model coupled to linear elasticity

$$\partial_t S = \ -c_H\left(W_S(S,\varepsilon) + \psi_2'(S) - \nu\Delta S\right)$$

and the degenerate mobility function

$$c_H(r) = |\nabla S|c(r) = |\nabla S|cr\,, \tag{3.6}$$

see [12], yields Eq. (3.5). Thus, the elastic hybrid model arises from the weighted gradient flow of the respective free energy and a function depending on the order parameter S induces the respective metric, see [32]. □

The elastic energy $W(S,\varepsilon)$ is defined by Eq. (2.3) and

$$\psi_2(S) = \frac{S(S-1)(S+0.06)(S-1.06)}{(\frac{1}{2}+0.06)^2} \tag{3.7}$$

is the double well function, given in [4], which is very close to $\psi(S)$ in Eq. (2.5), see Fig. 3.1. The slight difference between $\psi(S)$ and $\psi_2(S)$ is related to technical details of the proofs in [4].

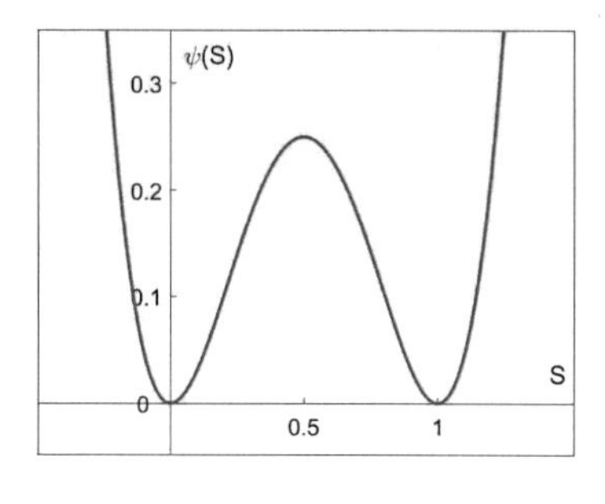

(a) $\psi(S)$ Allen-Cahn model

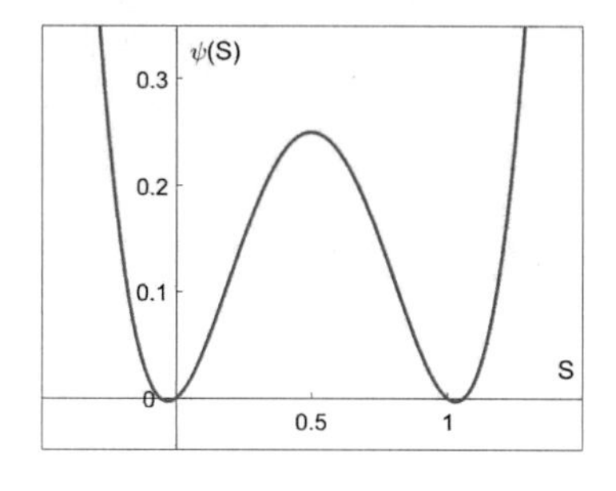

(b) $\psi_2(S)$ hybrid model

Figure 3.1: Double well potentials.

In the following, we will set $\psi_2(S)$ to $\psi(S)$, so Eq. (3.5) turns to

$$\partial_t S = -c\,(W_S(S,\varepsilon) + \psi'(S) - \nu\Delta S)|\nabla S|\,. \tag{3.8}$$

We choose the initial- and boundary conditions from Eq. (2.15).

Using Eq. (3.8), setting the parameter $\nu = 0$ and dropping the elasticity term W_S, gives the first order hyperbolic Hamilton-Jacobi type equation

$$\partial_t S = -c\,\psi'(S)|\nabla S|\,. \tag{3.9}$$

In [35], existence and uniqueness of Hamilton-Jacobi equations are examined by introducing the term $-\epsilon\Delta u^\epsilon$, corresponding to the Laplacian term in Eq. (3.8), denoted as the method of vanishing viscosity. In this context, the coefficient ν can be seen as a viscosity coefficient regularising the Hamilton-Jacobi equation.

Therefore, Eq. (3.8), representing a degenerative parabolic partial differential equation, shows properties of a hyperbolic Hamilton-Jacobi equation. For this reason, the authors in [4] named it hybrid model.

Asymptotic solutions In phase field modeling, asymptotic solution methods use inner and outer expansions for the unknowns and match them in the neighbourhood of $\Gamma(t)$. This technique leads to a sharp interface formulation of the respective phase field models, see [44], [47], [59], [9]. To explain the asymptotic approach for the hybrid model with respect to the original literature, the nomenclature, used in [4], is adopted in this paragraph.

Summary of the original literature Some aspects of the original literature [4], we will explain next, are very useful for the understanding of the numerical considerations in Chapter 5 and Chapter 7. Some more details are given in Appendix 1 and for the rest we refer to the proofs in [4].

In the notation of [4], the system of the hybrid phase field model coupled with equations of linear elasticity reads

$$-\mathrm{div}_x T = b\,, \tag{3.10}$$

$$T = D(\varepsilon(\nabla_x u) - \bar{\varepsilon} S)\,, \tag{3.11}$$

$$\partial_t S = -f(\psi_S(S, \varepsilon(\nabla_x u)) - \nu \Delta_x S)|\nabla_x S|\,. \tag{3.12}$$

The stress tensor is denoted by $T(x,t) \in \mathcal{S}^3$, the external time independent force by $b \in \mathbb{R}^3$, the linear elasticity tensor by $D : \mathcal{S}^3 \to \mathcal{S}^3$ (*in the general notation of the present work, D corresponds to* $\mathbb{C}$), the strain tensor by $\varepsilon(\nabla_x u) \in \mathcal{S}^3$, the displacements by $u(x,t) \in \mathbb{R}^3$, the eigenstrain by $\bar{\varepsilon} \in \mathcal{S}^3$ and the order parameter by $S(x,t) \in \mathbb{R}$. The function $f(r)$ describes the mobility, the function $\psi_S(S, \varepsilon) \in \mathbb{R}$ represents the sum $W_S(S,\varepsilon) + \psi_2'(S)$ as in Eq. (3.1) (*in the general notation of the present work, we have no corresponding term to* ψ_S) and $\nu \in \mathbb{R}^+$ is a small parameter.

The asymptotic solutions approaches $u^{(\nu)}(x,t)$ for $u(x,t)$, $S^{(\nu)}(x,t)$ for $S(x,t)$ and $T^{(\nu)}(x,t)$ for $T(x,t)$ are in [4] given as

$$u^{(\nu)}(x,t) = \phi(x,t) \sum_{i=0}^{1} \nu^{\frac{1+i}{2}} u_i(\eta, \zeta, t) + v(x,t), \tag{3.13}$$

$$S^{(\nu)}(x,t) = \phi(x,t) \sum_{i=0}^{1} \nu^{\frac{i}{2}} S_i(\eta, \zeta, t) + (1 - \phi(x,t))\, \hat{S}(x,t)\,, \tag{3.14}$$

$$T^{(\nu)}(x,t) = D\left(\varepsilon\left(\nabla u^{(\nu)}(x,t)\right) - \bar{\varepsilon} S^{(\nu)}(x,t)\right). \tag{3.15}$$

The Eqn. (3.13)-(3.15) are based on a local coordinate variable

$$\zeta = \frac{\xi}{\nu^{1/2}}, \tag{3.16}$$

defined by local coordinates (ξ, η) attached to the neighbourhood of the interface

$$\mathcal{U} = \{(\eta + n(t,\eta)\xi, t) \mid (\eta, t) \in \Gamma, \, |\xi| < \delta\} \subset \mathbb{R}^3 \times [t_1, t_2], \tag{3.17}$$

see Fig. 3.2. Γ denotes the sharp interface between the two material phases of $\Omega \subset \mathbb{R}^3$ in the notation of [4] (*in the general notation of the present work, Γ corresponds to $\hat{\Gamma}$*). The terms $u_i \in \mathbb{R}^3$ and $S_i \in \mathbb{R}$ are unknown functions to be determined, the function $\phi(x,t)$ indicates a point within or outside of $\mathcal{U}$ and $\hat{S}(x,t)$ is the order parameter of the respective phase (*in the general notation of the present work, we have $\hat{S}$ denoting the nodal values of S in the Finite Element formulation. This is very different from the meaning of $\hat{S}$ in [4]. Expressions for asymptotic solution terms are not defined in the present work.*). The displacement of the respective phase is denoted by $v(x,t)$. The inner and the outer expansion of the unknowns are matched in

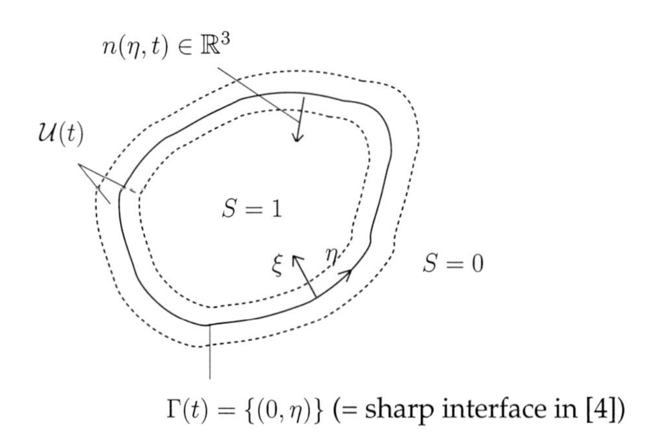

Figure 3.2: Local coordinate system in [4].

the neighbourhood region $\mathcal{U}(t)$ of the interface. For a diffuse interface, they must describe the order parameter $S \in (0,1)$ increasing smoothly and continuously within the neighbourhood domain and they must give the value $S(x,t) = 0$ for the first phase and the value $S(x,t) = 1$ for the second phase.

The approaches (3.13)–(3.15) are inserted into the system (3.10)–(3.12) and satisfy it asymptotically for $\nu \to 0$. This means that the residual estimates go to zero with $\nu \to 0$. This fact is proven by defining the normal velocity of the sharp interface as $s = -\partial_t \xi$, see Appendix 1 for more details and further definitions. The calculation leads to the normal sharp interface velocity in terms of

$$s = f(n \cdot [\hat{C}]n + \nu^{1/2} \omega_1 \kappa_\Gamma) \quad \text{with} \quad \omega_1 = \int_0^1 \sqrt{2\tilde{\psi}_1(\Theta)}\, d\Theta\,. \tag{3.18}$$

The modified double well potential $\tilde{\psi}_1$ (*in the general notation of the present work,* $\tilde{\psi}_1$ *corresponds to* ψ) is similar to Eq. (3.7) and will not be explained in detail, see [4] for the exact definition. The function f (*in the general notation of the present work,* f *corresponds to* f_2) is not primary linear. Anyhow, the authors in [4] restricted it to a linear case for their analytical considerations and their numerical examples. In the present work, this will be done in the same way. As explained above, the difference between the double well potentials of the Allen-Cahn model and the hybrid model is very small and thus we set $\tilde{\psi}_1 =: \psi$, see. Fig 3.1. The curvature of the sharp interface is denoted by κ_Γ.

Concluding we point out two important aspects of the proofs in [4]:

- The residuals of the system (3.10)–(3.12) with the inserted asymptotic approaches for the hybrid phase field model go to zero with $\nu \to 0$. The estimates of these residuals use the normal velocity definition of the sharp interface in terms of $s = f(n \cdot [\hat{C}]n + \nu^{1/2}\omega_1\kappa_k)$. Note that $f(r)$ in the original literature refers to $f_2(r)$ in the present work.
- The proofs for existence and the bounds of the asymptotic solutions for the hybrid model are based on the scaling of the local variable $\zeta = \frac{\xi}{\nu^{1/2}}$. This motivates the scaling employed in Chapter 5.4.

Especially the scaling allows a coarser grid at the interface for the hybrid model compared to the Allen-Cahn model. This fact leads to a numerical advantage and we will explain this aspect in Chapter 5.4 and Chapter 7.

4 Outline

In Part I we introduced the Allen-Cahn model and the hybrid model containing in particular a term of linear elasticity.

Part II describes phase field modeling without elasticity and we focus on the commonalities and the differences of the models without the elasticity term W_S. In Chapter 5, a general form of Allen-Cahn type phase field equations related to a non-elastic free energy is derived, including both models with the respective parameters. A general variational formulation is given and the energy decay property of the general model is shown.

In the second part of Chapter 5, we explain the error terms in the modelling presented. We distinguish between low and high interfacial energies determining the phase field evolution. Their magnitude influences the deviation of the numerical solutions from the expressions of the (sharp) interfacial velocities of the original literature. We present a relationship for the widths of the interfaces of both models that motivates the advantage of coarser mesh using the hybrid model in numerical implementations. We include the elastic terms within this error analysis, preparing the elastic simulations of Part III as well. This is an important part of the present work as it is the basis for the understanding of the results of the numerical implementation.

In Chapter 6, the general model without elasticity is discretised in time and space. The well-posedness of the discretised formulation and its energy decay property are discussed to confirm the numerical simulation results in the following chapter. Chapter 7 then gives one- and two-dimensional numerical examples with a focus on the width-development and the propagation velocity of the interfaces. Varying the parameters, we understand the differences between the two specific models. We compare some numerical results to available analytical solutions.

In Part III of this work, the general phase field model is expanded to linear elasticity. Chapter 8 explains the general formulation for the elastic phase field problem including the discussion of the energy decay and the well-posedness of the equations. Chapter 9 prepares the implementation by fully discretising the models. The proofs of well-posedness are partially transferred to the fully discretised formulation and we show the energy decay properties as well.

Part III concludes with examples of the numerical validation in Chapter 10. For numerical simulations with a simplified stress term adapted from [4], the results are compared to an analytical solution. Further examples are related to validated results in literature in the context of martensite transformations.

A difficulty in applying the hybrid model within very small scales is discussed in Part II and Part III. We analyse and explain this behaviour, which has not yet been published, and choose parameter values different from the Allen-Cahn model to reach comparable results for the martensite transformations using the hybrid model.

The work ends with some thoughts on further applications and open questions for future research.

Part II: System without elasticity

5 A general phase field model

In Part II and III we discuss discrete formulations of the phase field equations and examine the well-posedness of the stated problems. This ensures the reliability of the implementations and thus of the numerical results we will present.

To simplify our considerations, we restrict the numerical examples of Part II to phase field problems without coupled elastic material laws. For this reason, we preminilary drop the elastic term $W(S, \varepsilon)$.

Setting $W(S, \varepsilon) = 0$ and inserting $\alpha = \mu^{1/2}\lambda$, $\beta = \mu^{-1/2}$ for the Allen-Cahn model and $\alpha = \nu$, $\beta = 1$ for the hybrid model, the free energy Eq. (2.2) of the Allen-Cahn model and the free energy Eq. (3.4) of the hybrid model can be formulated as a general energy function in terms of

$$\bar{E}(S) = \int_{\Omega} \beta\psi(S) + \frac{\alpha}{2}|\nabla S|^2 \, dx \,. \tag{5.1}$$

As discussed in Chapter 2 and 3 and [22], solutions of the order parameter S follow the negative gradient of a respective free energy function in order to reduce the energy of the system.

The general phase field model is formulated as

$$h(|\nabla S|)\partial_t S = -\beta\psi'(S) + \alpha\Delta S \,, \quad \alpha \geq 0, \; \beta \geq 0 \,. \tag{5.2}$$

We define $h(q) = 1/c_A$ for the Allen-Cahn model, with c_A given by Eq. (2.12), and $h(q) = 1/q$ for the hybrid model, with $q = c_H$, given by Eq. (3.6), choosing $c = 1$.

To avoid dividing by zero, we regularise this term for the hybrid model by defining

$$h(|\nabla S|) := \begin{cases} \left(\sqrt{|\nabla S|^2 + \varepsilon_z}\right)^{-1}, \; 0 \leq \varepsilon_z \ll 1 & \text{hybrid model}\,, \\ \dfrac{(\mu\lambda)^{1/2}}{\tilde{c}} & \text{Allen-Cahn model}\,. \end{cases} \tag{5.3}$$

Thus, we can formulate

Assumption 2. $0 < \underline{h} \leq h(q) \leq \bar{h} < \infty \quad \forall q \geq 0.$

Remark 5.1. The function $h(q)$ stands as reciprocal of the mobility functions on the left-hand side of Eqn. (5.2). In the original literatur [4], $h(q) = q^{-1}$ multiplies the right-hand side like in Eq. (5.6) and $q^{-1} = |\nabla S| = 0$ indicates a constant phase outside the neighbourhood of the interface. The last property is somehow destroyed by the definition in Eq. (5.3), but since ε_z is very small, we get $h(q)^{-1} \approx 0$ in the pure phases modeled with the hybrid model.

The unusual representation with $h(|\nabla S|)$ on the left-hand side of Eq. (5.2) is used within the numerical implementation and therefore already introduced here.

Remark 5.2. We can write Eq. (5.2) as

$$\partial_t S = \alpha h(|\nabla S|)^{-1} \Delta S - \beta h(|\nabla S|)^{-1} \psi'(S) . \tag{5.4}$$

This is a quasilinear (degenerate) parabolic partial differential equation, since the coefficient function of the highest derivative of the unknown S depends on lower derivatives of S.

Inserting α, β and $h(|\nabla S|)$, we have the respective phase field equations

$$\partial_t S = -\frac{\tilde{c}}{(\mu\lambda)^{1/2}} \left(\frac{1}{\mu^{1/2}} \psi'(S) - \mu^{1/2} \lambda \Delta S \right) , \tag{5.5}$$

$$\partial_t S = -c \left(\psi'(S) - \nu \Delta S \right) |\nabla S| , \tag{5.6}$$

referring to the formulations (2.14) and (3.8).

Eq. (5.5) and Eq. (5.6) are supplemented by the initial- and boundary conditions given by Eq. (2.15).

To confirm the well-posedness of the discretised phase field formulation in Chapter 6, we need conditions on the term $\psi(S)$. Throughout this work we require the following

Assumption 3. *We assume that $\psi(S)$ is twice continuously differentiable with respect to S, $\psi(S) \geq 0$, $\psi(S) \geq C_{\psi,0}|S| - C_{\psi,1}$ and $|\psi''(S)| \leq C_{\psi,2}$ with constants $C_{\psi,i} \in \mathbb{R}_0^+$, $i = 0, 2 \ \ \forall S \in I_{a,b} = [a, b], \ a, b \in \mathbb{R}$.*

The assumption is evident for the definition of the double well potential given by Eq. (2.5) in $I = [0, 1]$ and an extendable to compact intervals $I_{a,b} \in [a, b], \ a, b \in \mathbb{R}$.

To ensure the upper limit, the double well potential must be changed to a function that tends towards a constant maximum outside of $I_{a,b}$. Modified potentials can be found in literature and will not be discussed in the present work.

5.1 Variational formulation

In contrast to equation Eq. (5.5), Eq. (5.6) generally does not have a classical solution. An approach for a numerical treatment of this type of (degenerate) quasilinear parabolic differential equation is given by a variational formulation.

Therefore, we multiply Eq. (5.2) with a test function $v_S \in H_0^1(\Omega)$ and integrate over the domain Ω to the result

$$\int_\Omega h(|\nabla S|)\partial_t S v_S \, dx = -\beta \int_\Omega \psi'(S) v_S \, dx + \alpha \int_\Omega (\Delta S) v_S \, dx \, .$$

Integrating by parts and using homogeneous boundary conditions, we obtain

$$\int_\Omega h(|\nabla S|)\partial_t S v_S \, dx + \beta \int_\Omega \psi'(S) v_S \, dx + \alpha \int_\Omega \nabla S \cdot \nabla v_S \, dx = 0 \, .$$

This can be rewritten as

$$\langle h\left(|\nabla S|\right) \partial_t S, v_S\rangle + \beta\langle \psi'(S), v_S\rangle + \alpha\langle \nabla S, \nabla v_S\rangle = 0 \, , \quad \forall v_S \in H_0^1(\Omega), \quad t > 0 \, , \tag{5.7}$$

with the L^2-scalar product $\langle v, w\rangle = \int_\Omega vw \, dx$.

5.2 Energy decay

For physical reasons, the value of the free energy function of the respective phase field model should decay in time. We want to verify this in the following.

Lemma 3 (*Energy decay of the general model*)**.** *Let S be a sufficiently smooth solution of Eq. (5.2) and Eq. (2.15). Further assume $\partial_t S \in H_0^1(\Omega)$. Then the free energy (5.1) satisfies*

$$\frac{d}{dt}\bar{E}(S(t)) = -\int_\Omega h\left(|\nabla S(t)|\right) \partial_t S^2 \, dx \leq 0 \, . \tag{5.8}$$

Proof. Taking the derivative of Eq. (5.1) with respect to time t we have

$$\frac{d}{dt}\bar{E}(S(t)) = \int_\Omega \beta\psi'(S(t))\partial_t S + \alpha\nabla S(t) \cdot \nabla(\partial_t S)\, dx\,.$$

Now we choose $v_S = \partial_t S$ as a test function for Eq. (5.7) yielding

$$\langle h\left(|\nabla S(t)|\right)\partial_t S, \partial_t S\rangle = -\langle \beta\psi'(S(t)), \partial_t S\rangle - \langle \alpha\nabla S(t), \nabla\partial_t S\rangle$$

and calculate

$$\frac{d}{dt}\int_\Omega \bar{E}(S(t)) = -\langle h\left(|\nabla S(t)|\right)\partial_t S, \partial_t S\rangle = -\int_\Omega h\left(|\nabla S(t)|\right)(\partial_t S)^2\, dx \leq 0\,.$$

□

5.3 Correlation

So far we considered a general phase field model including the Allen-Cahn model and the hybrid model as special cases. Remarkable is the definition of $h(|\nabla S(x,t)|)$ by Eq. (5.3) for the hybrid model changing the character of the equation compared to the constant mobility given for the Allen-Cahn equation. Numerically we can treat both equations in the same way using the general formulation Eq. (5.2) although their analytic type of partial differential equation differs. We will explain the numerical implementation of the general form in more detail in Chapter 6 and Appendix 4. Before that, we want to study an aspect that will play an important role in the numerical analysis.

The authors of [3] and [4] state that if both models are acting on the same topology with the same (physical) parameters and equal boundary conditions, their convergence to a sharp interface solution would be approximately the same. Additionally, they argue that the numerical simulation effort for the hybrid model could be reduced in a significant way due to a coarser mesh especially inside the neighbourhood of the interface.

Anyhow, since the values of the parameters μ, ν and λ as well as $\tilde{c}$ and c in Eq. (5.5) and Eq. (5.6) have a distinguished meaning, as explained in Chapter 2 and 3, we could ask if the two models behave really in the same way. To consider this, we will examine the commonalities and the differences of both models in the following.

Remark 5.3. To achieve sufficient generality also for the third part of this work, we include the elasticity term W_S and the corresponding term $n \cdot [\hat{C}]n$ in the expression of the normal interface velocity. All expressions given in the following are valid for the non-elastic case as well, setting both terms to zero.

In [2] and [3], one-dimensional existence proofs are derived, yielding an expression for the normal interface velocity given as

$$s_{AC}^{el} = g\left(n \cdot [\hat{C}]n + \lambda^{1/2} c_1 \kappa_\Gamma + O(\mu^{1/2})\right) \tag{5.9}$$

for the Allen-Cahn model and

$$s_H^{el} = f_2\left(n \cdot [\hat{C}]n + \nu^{1/2} \omega_1 \kappa_\Gamma\right) \tag{5.10}$$

for the hybrid model. They refer to the formulations Eq. (2.14) and Eq. (3.8), containing the elasticity term W_S which is presented by the term $n \cdot [\hat{C}]n$.

The functions in front of the brackets on the right-hand side can in general be non-linear and we will specify them in the following. We want to indicate that the expression $O(\mu^{1/2})$ in Eq. (5.9) was not part of the interface velocity in [4] but in the later publication [3].

Eq. (5.10) corresponds to Eq. (3.18) in Chapter 3. Note that we name the function $f(r)$ of the original literatur $f_2(r)$ in the present work, to distinguish it from the function $f_1(r)$ related to the Allen-Cahn model. As in the original literature, the function $f_2(r)$ in Eq. (5.10) is concordant with the one in Eq. (3.1).

The connection between the functions $g(r) = gr,\ g \in \mathbb{R}^+$ and $\tilde{c}(r) = \tilde{c}r,\ \tilde{c} \in \mathbb{R}^+$ in Eq. (2.14) and in Eq. (5.9) is explained further below. The definition of c_1 in Eq. (2.17) is close to ω_1 in Eq. (3.18), since the double well potentials of the Allen-Cahn model and the hybrid model differ only slightly, see Fig. 3.1.

Omitting the elasticity term W_S, the term $n \cdot [\hat{C}]n$ vanishes and Eq. (5.9) and Eq. (5.10) read

$$s_\mu = g\left(\lambda^{1/2} c_1 \kappa_\Gamma + O(\mu^{1/2})\right), \tag{5.11}$$

$$s_\nu = f_2\left(\nu^{1/2} \omega_1 \kappa_\Gamma\right). \tag{5.12}$$

These expressions can be interpreted in two ways. Since Eq. (5.11) and Eq. (5.12) both contain the curvature term κ_Γ, related to an interface energy, they can be seen as descriptions of a curvature dependent driving force. On the other hand, if this interface energy is very small compared to the rest of the total free energy, as, e.g. in martensite transformations, the expressions in Eq. (5.11) and Eq. (5.12) can be seen as error terms. We will examine this second aspect in the following chapter.

5.4 Error terms

To compare the function $g(r) = gr$ in Eq. (5.9) with the function $f_2(r)$ in Eq. (5.10), we state that $f_2(r) = cr$ as in (3.8). The conformity of f_2 in Eq. (5.10) and Eq. (3.1) is valid for a non-linear function $f_2(r)$ as well. This more general case will not be discussed in the present work.

The fact that the constant functions $g(r)$ in Eq. (5.9) and $f_1(r)$ in Eq. (2.13) are connected by an integral operator is stated in [4] and we will have a short view on the involved terms for the linear case.

Lemma 4 (*Integral operator*). *The functions $g(r)$ in Eq. (5.9) and $f_1(r)$ in Eq. (2.13) are connected by an integral operator*

$$g^{-1}(r) = \int_0^1 f_1^{-1}\left(r\sqrt{2\psi(\zeta)}\right)\, d\zeta \tag{5.13}$$

and lead for the linear relation, $g(r) = gr$, to the kinetic expression Eq. (2.16) of the Allen-Cahn model.

Proof. With $g(r) = g\,r$, Eq. (2.11) and c_1, defined by Eq. (2.17), we calculate

$$\begin{aligned} g^{-1}(r) &= \int_0^1 \tilde{c}^{-1}\left(r\sqrt{2\psi(\zeta)}\right)\, d\zeta \\ &= \tilde{c}^{-1}(r)\int_0^1 \sqrt{2\psi(\zeta)}\, d\zeta = c_1\,\tilde{c}^{-1}(r) = \left(\frac{\tilde{c}}{c_1}\right)^{-1}(r)\,, \end{aligned}$$

$$\text{thus,}\quad g(r) = \frac{\tilde{c}}{c_1}(r)\,.$$

Inserting $r = n\cdot[\hat{C}]n + \lambda^{1/2}c_1\kappa_\Gamma + O(\nu^{1/2})$ and neglecting the error term by setting $O(\nu^{1/2}) = 0$, leads to

$$g(r) = \frac{\tilde{c}}{c_1}\left(n\cdot[\hat{C}]n + \lambda^{1/2}c_1\kappa_\Gamma\right)\,.$$

With $s = g(r)$, the normal interface velocity of the Allen-Cahn model Eq. (2.16) is given.

□

For non-linear functions $g(r)$ and $f_1(r)$ the lemma holds also as shown in [4].

Remark 5.4. The third term on the right-hand side of Eq. (5.9) is not provided in [4], where asymptotic solutions of second order for the hybrid model are compared to asymptotic solutions of first order for the Allen-Cahn model. In [3], asymptotic solutions of second order are discussed for the Allen-Cahn model also and lead to the additional term $O\left(\mu^{1/2}\right)$ in Eq. (5.9). Numerical examples in [4] therefore do not consider the $O\left(\mu^{1/2}\right)$-term that however is part of the model error.

Including the error-term $O\left(\mu^{1/2}\right)$ of the second-order asymptotic for the Allen-Cahn model, we compare the normal interface velocities Eq. (5.9) and Eq. (5.10) with linear mobility functions in terms of

$$s_1 = \frac{\tilde{c}}{c_1}\left(n\cdot[\hat{C}]n + \lambda^{1/2}c_1\kappa_\Gamma + O\left(\mu^{1/2}\right)\right), \tag{5.14}$$

$$s_2 = c\left(n\cdot[\hat{C}]n + \nu^{1/2}\omega_1\kappa_\Gamma\right), \tag{5.15}$$

including the elastic terms $n\cdot[\hat{C}]n$, considering the extension in Part III.

For equal double well potentials we have $c_1 = \omega_1$ and the curvature terms are equal if we choose the same values for λ and ν. For simplicity, we define $c = 1$ and set $\tilde{c} = c_1$. With these definitions, the differential equations for the phase field models Eq. (2.14) and Eq. (3.8) read

$$\partial_t S = -\frac{c_1}{(\mu\lambda)^{1/2}}\left(W_S(S,\varepsilon) + \frac{1}{\mu^{1/2}}\psi'(S) - \mu^{1/2}\lambda\Delta S\right), \tag{5.16}$$

$$\partial_t S = -\left(W_S(S,\varepsilon) + \psi'(S) - \nu\Delta S\right)|\nabla S|. \tag{5.17}$$

Table 5.2 summarises the previous considerations including the elastic term that will be dropped in the following.

The function $f_1(r)$ represents the possibly non-linear formulation of the mobility function of the Allen-Cahn model that we do not adress in the present work. Anyhow, we include it in this table for a conforming description of both models.

In Eq. (5.14) and Eq. (5.15), the constant terms $\lambda^{1/2}c_1$ and $\nu^{1/2}\omega_1$ stand in front of the curvature and therefore λ and ν can be interpreted as interface energy parameters. At once, the parameter ν is an asymptotic solution parameter, see Chapter 3. Thus, setting $\lambda = \nu$ implies that the size of the interface energy parameter λ determines the convergence behaviour against the sharp interface solution of the hybrid model.

Table 5.2: Summary of the model parameters.

Allen-Cahn model	hybrid model
$\partial_t S = -\frac{1}{(\mu\lambda)^{1/2}} f_1(W_S(S,\varepsilon) + \frac{1}{\mu^{1/2}}\psi'(S) - \mu^{1/2}\lambda\Delta S)$	$\partial_t S = -f_2(W_S(S,\varepsilon) + \psi'(S) - \nu\Delta S)\lvert\nabla S\rvert$
$s^{el}_{AC} = g\left(n\cdot[\hat{C}]n + \lambda^{1/2}c_1\kappa_\Gamma + O(\mu^{1/2})\right)$	$s^{el}_{H} = f_2\left(n\cdot[\hat{C}]n + \nu^{1/2}\omega_1\kappa_\Gamma\right)$
$f_1(r) = \tilde{c}r,\ g(r) = \frac{\tilde{c}}{c_1}r, c_1 \in \mathbb{R}^+,\ \tilde{c} \in \mathbb{R}^+$	$f_2(r) = cr,\ c \in \mathbb{R}^+$
choose $\tilde{c} = c_1 c$.	choose c.

Conclusion:

- The choice of $f_1(r) = \tilde{c}r$ determines $g(r) = gr$ for the Allen-Cahn model.
- Comparing the hybrid model to the Allen-Cahn model implies to set $f_2(r) = g(r)$ and thus, defines the mobility function for the hybrid model.

In case of very small interface energies ($\lambda \to 0$ and $\nu = \lambda$) we identify

$$\hat{s} = n\cdot[\hat{C}]n \tag{5.18}$$

as the driving force due to, e.g. elastic deformations and we regard the remaining terms of Eq. (5.14) and Eq. (5.15) as error terms

$$s_{1\mu} = s_1 - \hat{s}\,, \tag{5.19}$$
$$s_{2\nu} = s_2 - \hat{s}\,. \tag{5.20}$$

The interfacial velocities given by Eq. (5.9) and Eq. (5.10) have to be supplemented by an additional model error term originating from the residual estimates in [4]. We call this $e^{\mu}_{res} = O(\mu^{1/2})$ for the Allen-Cahn model and $e^{\nu}_{res} = O(\nu)$ for the hybrid model and define

$$s_{1_{err}} = s_{1\mu} + e^{\mu}_{res}\,, \tag{5.21}$$
$$s_{2_{err}} = s_{2\nu} + e^{\nu}_{res} \tag{5.22}$$

and with $c = 1$ and $\tilde{c} = c_1$ (see Table 5.2) we have

$$s_{1_{err}} = \lambda^{1/2} c_1 \kappa_\Gamma + O(\mu^{1/2}) + O(\mu^{1/2}), \tag{5.23}$$
$$s_{2_{err}} = \nu^{1/2} \omega_1 \kappa_\Gamma + O(\nu)\,. \tag{5.24}$$

The order of the error terms assuming negligible interfacial energy are hence given by

$$s_{1_{err}} = O\left(\lambda^{1/2}\right) + O\left(\mu^{1/2}\right)\,, \tag{5.25}$$
$$s_{2_{err}} = O\left(\nu^{1/2}\right)\,. \tag{5.26}$$

So we have to distinguish two situations. In case of a non-negligible interfacial energy we set $\nu = \lambda$ and the first terms on the right-hand sides of Eq. (5.25) and Eq. (5.26) stand for the curvature influence on the driving force at the interface. They are represented by the second terms on the right-hand sides of Eq. (5.14) and Eq. (5.15).

In this situation of influential interfacial energy, the asymptotic parameter μ can be chosen sufficiently small for the Allen-Cahn model, minimising the error. Since the normal velocities of the sharp interfaces depend on the sizes of λ and ν, the $O-$terms in Eq. (5.23) and Eq. (5.24) additionally illustrate that the choice of a large $\lambda = \nu$ (> 1) increases the error of the hybrid model quadratically compared to the influence of the curvature term. Thus, the Allen-Cahn model is suited better for significant interfacial energies. We can even say that the hybrid model is quite unsuitable in this case.

In the other case, we consider a problem with very small interface energy. A small value for λ however relates to a small value for ν. Choosing $\lambda = \nu$ very small, the sharp interface solution for the hybrid model converges well, recognisable in Eq. (5.24) and Eq. (5.26). To achieve a comparable convergency for the Allen-Cahn model, the parameter μ has to be chosen in the order of λ or smaller, see Eq. (5.23) and Eq. (5.25). If we choose at least $\mu = \lambda$, the scaling of the diffuse interface shows a disadvantage. We will explain this fact in the following using the notation of the original literature.

Scaling in terms of the original literature. We recall the asymptotic approach for the hybrid model in Chapter 3 in the notation of [4],

$$S^{(\nu)}(x,t) = \phi(x,t) \sum_{i=0}^{1} \nu^{\frac{i}{2}} S_i \left(\eta, \frac{\xi}{\nu^{1/2}}, t \right) + (1 - \phi(x,t))\, \hat{S}(x,t) \tag{5.27}$$

and compare it to the asymptotic approach for the Allen-Cahn model in [3],

$$S^{(\mu)}(x,t) = S_1^{(\mu)}(x,t)\, \phi_{\mu\lambda}(x,t) + S_2^{(\mu)}(x,t)\, \left(1 - \phi_{\mu\lambda}(x,t)\right) .$$

The function $\phi_{\mu\lambda}(x,t)$ is a C_0^∞-function that is equal to 1 for the phase with $S = 1$, transits smoothly from 1 to 0 within the interface region and vanishes for the phase with $S = 0$. Thus, the asymptotic solution terms $S_1^{(\mu)}$ and $S_2^{(\mu)}$ are, depending on the respective phase, switched on and off and weighted within the neighbourhood $\Gamma(t)$. The outer expansion term is defined by

$$S_2^{(\mu)}(x,t) = \hat{S}(x,t) + \mu^{1/2} \tilde{S}_1(x,t) + \mu \tilde{S}_2(x,t) + \mu^{2/3} \tilde{S}_3(x,t)$$

with the unknown functions $\tilde{S}_i$, $i = 1, 3$. The function $\hat{S}$ is constant 0 or 1 with a jump at the sharp interface. Note that the third term on the right-hand side with $\tilde{S}_3$ is not part of the outer expansion in [4].

The inner expansions are in [3] given as

$$S_1^{(\mu)}(x,t) = S_0\left(\frac{\xi}{(\mu\lambda)^{1/2}}\right) + \mu^{1/2} S_1\left(\eta, \frac{\xi}{(\mu\lambda)^{1/2}}, t\right) + \mu S_2\left(\eta, \frac{\xi}{(\mu\lambda)^{1/2}}, t\right) \tag{5.28}$$

with the unknown functions S_i, $i = 0, 2$. Here, also the last term including S_2 on the right-hand side is not given in [4]. The additional terms with $\tilde{S}_3$ and S_2 in [3] lead to the additional term $(\mu^{1/2})$ in Eq. (5.9) that is part of the normal interface velocity in [3] but not in [4]. Comparing Eq. (5.27) and Eq. (5.28), we recognise the author's argument that the hybrid model scales with $\nu^{1/2}$ and the Allen-Cahn model scales with $(\mu\lambda)^{1/2}$. This is the reason why the numerical widths of the interfaces relate like

$$re_A = \frac{\text{width}_{AC}}{\text{width}_H} = \frac{(\mu\lambda)^{1/2}}{\nu^{1/2}} \tag{5.29}$$

for the same precision of the results. For modeling a system with low interfacial energy we choose $\mu \leq \lambda$ to control the error term for the Allen-Cahn model and $\lambda = \nu$ to compare it to the hybrid model. For the limit case $\mu = \lambda$ the width relation for the interfaces thus yields

$$re_B = \frac{\text{width}_{AC}}{\text{width}_H} = \nu^{1/2} . \tag{5.30}$$

Thus, in the case of low interfacial energy, the hybrid model is significantly better suited if we want a coarser mesh, because a relation $\nu^{1/2} = 0.1$ means that we have a ten times wider diffuse interface of the hybrid model compared to the Allen-Cahn model for similar results. This implies a better numerical resolution of the interface area.

Table 5.3 is an extension of Table 5.2 and considers the different error behaviour of both models depending on the interfacial energy situation.

In summary we could say that in the case of a rather small interfacial energy problem, the choice $\mu = \lambda$ for the Allen-Cahn model might be motivated by Eq. (5.25): μ should not increase the error (for $\mu > \lambda$) but should also not decrease the smooth interface width (for $\mu < \lambda$) related to the coordinate $\xi(S \to 1)$ in Fig. 3.2. For that reason, $\mu = \lambda$ might be a reasonable choice for this case.

Using, e.g. a two-dimensional Finite Element method requires a fine resolution of the transition zone and the choice of the best compromise concerning the grid spacing for the Allen-Cahn model and for the hybrid model as well.

In [3], the authors stated the conflict between low numerical effort and good convergence behaviour as an optimisation problem for the total model error of the Allen-Cahn model ε_{L2}, defined by the interface width parameter $B = (\mu\lambda)^{1/2}$, the energy parameter $E = \lambda^{1/2}$ and the error parameter for the normal interface velocity $F = \mu^{1/2}$, related to Eq. (5.25). Further details are given in the original literature.

Numerical examples will be given in Chapter 7.

Table 5.3: Summary for the functions f_1, f_2, interfacial energies and error terms

Allen-Cahn model	hybrid model
$\partial_t S = -\frac{1}{(\mu\lambda)^{1/2}} f_1(W_S(S,\varepsilon) + \frac{1}{\mu^{1/2}}\psi'(S) - \mu^{1/2}\lambda\Delta S)$	$\partial_t S = -f_2(W_S(S,\varepsilon) + \psi'(S) - \nu\Delta S)\lvert\nabla S\rvert$
$s^{el}_{AC} = g\left(n\cdot[\hat{C}]n + \lambda^{1/2}c_1\kappa_\Gamma + O(\mu^{1/2})\right)$	$s^{el}_{H} = f_2\left(n\cdot[\hat{C}]n + \nu^{1/2}\omega_1\kappa_\Gamma\right)$
$f_1(r) = \tilde{c}\,r,\ g(r) = \frac{\tilde{c}}{c_1}r, c_1 \in \mathbb{R}^+, \tilde{c} \in \mathbb{R}^+$	$f_2(r) = c\,r,\ c \in \mathbb{R}^+$
choose $\tilde{c} = c_1 c$.	choose c.
high interfac. energy $\rightarrow$ choose λ large. error $\rightarrow s_{err,AC} = O\left(\mu^{1/2}\right)$ interface width $\rightarrow f(\mu)$	high interfac. energy $\rightarrow$ choose ν large. error $\rightarrow s_{err,H} = O\left(\nu\right)$ interface width $\rightarrow f(\nu)$
low interfac. energy $\rightarrow$ choose λ small. error $\rightarrow s_{err,AC} = O\left(\lambda^{1/2}\right) + O\left(\mu^{1/2}\right)$ interface width $\rightarrow f(\mu,\lambda)$ with $\mu \leq \lambda$	low interfac. energy $\rightarrow$ choose ν small. error $\rightarrow s_{err,H} = O\left(\nu^{1/2}\right)$ interface width $\rightarrow f(\nu)$

Conclusion: for a negligible interfacial energy, the numerical advantage of the hybrid model is high.

- If the curvature term can be neglected, we can make ν small and thereby minimise the error term using the hybrid model.
- Minimising the error and the curvature term by choosing $\lambda = \mu = \nu \ll 1$, the numerical effort increases for the Allen-Cahn model compared to the hybrid model.

6 Discretisation of the general model

Discretisation in space. The phase field Eq. (5.2) is implemented with the Finite Element method using MATLAB. We define shape functions $N_I(x)$ representing the basis of the finite dimensional subspaces $V_h \subset H^1(\Omega)$ and $V_{h,0} \subset H_0^1(\Omega)$, $\hat{S}_I$ for the nodal unknowns and a semi-discretised order parameter

$$S_h(x) = \sum_I^N N_I(x)\hat{S}_I \,.$$

The index 0 considers homogeneous Dirichlet boundary conditions, see Appendix 2 - 4 for details of the Finite Element implementation.

With these preliminary remarks we consider the semi-discrete phase field equation. We begin with the

Problem (*Variational semi-discrete formulation*). Let $S_{h,0} \in V_{h,0}$ be given. Find $S_h \in C^1([0,T], V_{h,0})$ such that

$$\langle h\left(|\nabla S_h|\right) S_{h,t}, v_{hS}\rangle + \beta\langle \psi'(S_h), v_{hS}\rangle + \alpha\langle \nabla S_h, \nabla v_{hS}\rangle = 0 \tag{6.1}$$

for all discrete test functions $v_{hS} \in V_{h,0}$ and $t > 0$.

With the semi-discrete energy function

$$\bar{E}(S_h) = \int_\Omega \beta\psi(S_h) + \frac{\alpha}{2}|\nabla S_h|^2 \, dx \tag{6.2}$$

the energy decay can be shown along the lines of the above given proof for the general continuous model.

Lemma 5 (*Energy decay for the semi-discretisation formulation*). *Let $S_h \in C^1([0,T]; V_{h,0})$ be a solution of Eq. (6.1). Then the free energy function satisfies the condition*

$$\frac{d}{dt}\bar{E}(S_h) = -\int_\Omega h\left(|\nabla S_h|\right) S_{h,t}^2 \, dx \leq 0 \,. \tag{6.3}$$

Proof. The semi-discretised energy function (6.2) can be differentiated with respect to the time t leading to

$$\frac{d}{dt}\bar{E}(S_h) = \int_\Omega \beta\psi'(S_h)S_{h,t} + \alpha\nabla S_h \cdot \nabla(S_{h,t}) \, dx \,. \tag{6.4}$$

Inserting $v_{hS} = S_{h,t}$ as a test function into Eq. (6.1) yields

$$\langle h\left(|\nabla S_h|\right) S_{h,t}, S_{h,t}\rangle = -\langle \beta\psi'(S_h), S_{h,t}\rangle - \langle \alpha \nabla S_h, \nabla S_{h,t}\rangle \tag{6.5}$$

and the combination of Eq. (6.5) and Eq. (6.4) gives the result

$$\frac{d}{dt}\bar{E}(S_h) = -\langle h\left(|\nabla S_h|\right) S_{h,t}, S_{h,t}\rangle \;\; = -\int_\Omega h\left(|\nabla S_h|\right) S_{h,t}^2 \, dx \leq 0\,.$$

□

Next, we examine the well-posedness of the semi-discretised phase field formulation Eq. (6.1) in terms of

Lemma 6 (*Existence and uniqueness of the semi-discrete formulation*)**.** *Let $S_{h,0} \in V_{h,0}$ be given. Then we have an existing unique solution $S_h \in C^1([0,1]; V_{h,0})$ of Eq. (6.1).*

Proof. We can restate Eq. (6.1) to

$$\tilde{H}(S_h)S_{h,t} + \tilde{F}(S_h) + \tilde{K}S_h = 0, \quad S_h(0) = S_{h,0}\,. \tag{6.6}$$

With the Assumptions 2 and 3 we have Lipschitz continuity of the functions $\tilde{K}$ and $\tilde{F}(\cdot)$ and an inverse for $\tilde{H}(S_h)$. Thus, the Picard-Lindelöf theorem guarantees a local unique solution of

$$S_{h,t} = -\tilde{H}(S_h)^{-1}(\tilde{F}(S_h) + \tilde{K}S_h)\,. \tag{6.7}$$

Eq. (6.7) is an autonomous differential equation. For this type of differential equation the boundedness of S_h allows to transfer the existence of a local unique solution to the existence of a global unique solution, see [51]. So we have to show the boundedness of S_h.

With the belonging semi-discrete energy (6.2) and the energy decay property (6.3) we can find a constant

$$C' := \bar{E}(S_{0,h}) \geq \bar{E}(S_h) = \beta||\psi(S_h)||_{L^1(\Omega)} + \frac{\alpha}{2}||\nabla S_h||^2_{L^2(\Omega)}\,. \tag{6.8}$$

With Assumption 3 we get

$$C' \geq \beta C_{\psi,0}||S_h||_{L^1(\Omega)} + \frac{\alpha}{2}||\nabla S_h||^2_{L^2(\Omega)}$$

and we can find a constant

$$C'' \geq ||\nabla S_h||_{L^2(\Omega)} . \tag{6.9}$$

With the definition of the Sobolev norm

$$||S_h||_{H^1(\Omega)} = \left(||S_h||^2_{L^2(\Omega)} + ||\nabla S_h||^2_{L^2(\Omega)}\right)^{1/2} \tag{6.10}$$

and the Poincaré inequality, using homogeneous boundary conditions, see [81], we have

$$||S_h||_{L^2(\Omega)} \leq C_P||\nabla S_h||_{L^2(\Omega)} . \tag{6.11}$$

So we can follow

$$\begin{aligned} ||S_h||_{H^1(\Omega)} &\leq \left(\tilde{C}_P||\nabla S_h||^2_{L^2(\Omega)} + ||\nabla S_h||^2_{L^2(\Omega)}\right)^{1/2} \\ &\leq \left((1+\tilde{C}_P)||\nabla S_h||^2_{L^2(\Omega)}\right)^{1/2} \\ &\leq \tilde{\tilde{C}}_P||\nabla S_h||_{L^2(\Omega)} \end{aligned}$$

and with the estimate (6.9) we see that

$$||S_h||_{H^1(\Omega)} \leq C''' . \tag{6.12}$$

According to this result and with the local existence from above, the lemma is proven. □

For further notes, see [22].

Discretisation in time. For the proof of uniqueness of the fully discretised formulation of problem (6.1) we need to approximate the derivative of the double well potential $\psi'(S_h)$ by the difference quotient

$$D(S_h^n, S_h^{n-1}) := \begin{cases} \dfrac{\psi(S_h^n) - \psi(S_h^{n-1})}{S_h^n - S_h^{n-1}} & \text{for } S_h^n \neq S_h^{n-1} , \\ \psi'(S_h^n) & \text{for } S_h^n = S_h^{n-1} . \end{cases} \tag{6.13}$$

The discrete time derivative of the order parameter $S_{h,t}$ in Eq. (6.1) is given by the backward difference quotient at time $t^n = n\tau$ in terms of

$$\bar{\partial}_\tau S_h^n = \frac{S_h^n - S_h^{n-1}}{\tau}. \tag{6.14}$$

We restate Eq. (6.1) with $h(|\nabla S_h^{n-1}|)$ instead of $h(|\nabla S_h^n|)$ and formulate the

Problem (*Variational fully-discrete formulation*). Let $S_{h,0} \in V_{h,0}$ be given. For any $n \geq 1$ find $S_h^n \in V_{h,0}$ such that

$$\langle h(|\nabla S_h^{n-1}|)\frac{S_h^n - S_h^{n-1}}{\tau}, v_{hS}\rangle + \beta\langle D(S_h^n, S_h^{n-1}), v_{hS}\rangle + \alpha\langle \nabla S_h^n, \nabla v_{hS}\rangle = 0 \tag{6.15}$$

for all discrete test functions $v_{hS} \in V_{h,0}$.

To prove the physical plausibility, we show the energy decay property for the fully-discrete scheme.

We define

$$\bar{\partial}_\tau Y_h^n = \frac{1}{\tau}(Y_h^n - Y_h^{n-1}). \tag{6.16}$$

Lemma 7 (*Energy decay for the fully discretised general model*). *Let $S_h^n \in V_{h,0}$ $\forall t$ be a solution of Eq. (6.15). Then the free energy satisfies the condition*

$$\bar{\partial}_\tau \bar{E}(S_h^n) = -\int_\Omega h(|\nabla S_h^{n-1}|)|\bar{\partial}_\tau S_h^n|^2\,dx - \tau\frac{\alpha}{2}||\bar{\partial}_\tau \nabla S_h^n||^2_{L^2(\Omega)} \leq 0. \tag{6.17}$$

Proof. From the semi-discretised free energy function (6.2) we calculate

$$\begin{aligned}
\bar{E}(S_h^n) - \bar{E}(S_h^{n-1}) &= \beta\langle\psi(S_h^n) - \psi(S_h^{n-1}), 1\rangle + \frac{\alpha}{2}\left(\langle\nabla S_h^n, \nabla S_h^n\rangle - \langle\nabla S_h^{n-1}, \nabla S_h^{n-1}\rangle\right) \\
&= \beta\langle\frac{\psi(S_h^n) - \psi(S_h^{n-1})}{S_h^n - S_h^{n-1}}, S_h^n - S_h^{n-1}\rangle + \alpha\langle\nabla S_h^n, \nabla S_h^n - \nabla S_h^{n-1}\rangle - \frac{\alpha}{2}||\nabla S_h^n - \nabla S_h^{n-1}||^2_{L_2(\Omega)}.
\end{aligned}$$

For the last step we used $\frac{1}{2}(a^2 - b^2) = a(a-b) - \frac{1}{2}(a-b)^2$ for $a = \nabla S_h^n$ and $b = \nabla S_h^{n-1}$.

Division by τ then yields

$$\begin{aligned}
\bar{\partial}_\tau \bar{E}(S_h^n) &= \frac{\beta}{\tau}\langle\frac{\psi(S_h^n) - \psi(S_h^{n-1})}{S_h^n - S_h^{n-1}}, S_h^n - S_h^{n-1}\rangle \\
&+ \frac{\alpha}{\tau}\langle\nabla S_h^n, \nabla S_h^n - \nabla S_h^{n-1}\rangle - \frac{\alpha}{2\tau}||\nabla S_h^n - \nabla S_h^{n-1}||^2_{L_2(\Omega)}.
\end{aligned} \tag{6.18}$$

Testing with $v_{hS} = (S_h^n - S_h^{n-1})/\tau$, Eq. (6.15) reads

$$\int_\Omega h(|\nabla S_h^{n-1}|)|\bar{\partial}_\tau S_h^n|^2 \, dx = -\frac{\beta}{\tau}\langle\frac{\psi(S_h^n) - \psi(S_h^{n-1})}{S_h^n - S_h^{n-1}}, S_h^n - S_h^{n-1}\rangle - \frac{\alpha}{\tau}\langle\nabla S_h^n, \nabla S_h^n - \nabla S_h^{n-1}\rangle$$

and insertion into Eq. (6.18) leads to

$$\bar{\partial}_\tau E(S_h^n) = -\int_\Omega h(|\nabla S_h^{n-1}|)|\bar{\partial}_\tau S_h^n|^2 \, dx - \tau\frac{\alpha}{2}||\bar{\partial}_\tau \nabla S_h^n||^2_{L^2(\Omega)} \leq 0\,. \tag{6.19}$$

□

Now we will look on the well-posedness of the fully-discrete phase field equation to confirm the numerical implementation.

Lemma 8 (*Existence and uniqueness of the fully discrete formulation*)**.** *Let $S_h^{n-1} \in V_{h,0}$ be given. Then Eq. (6.15) has a unique solution $S_h^n \in V_{h,0}$ for all time steps $0 < \tau < \tau_0$ with τ_0 small enough.*

Proof. We use the lemma from [35], known as the lemma of "Zeros of a vector field", which states that for a continuous function $v : \mathbb{R}^N \to \mathbb{R}^N$, satisfying

$$v(x) \cdot x \geq 0 \quad \text{if} \quad |x| = r \quad \text{for some} \quad r > 0\,, \tag{6.20}$$

there exists a point $x \in B_r(0)$ such that $v(x) = 0$.

To adapt our problem to the requirements of the lemma, we use that we can express the discrete order parameter by

$$S_h(x) = \sum_{I=1}^{N} N_I(x)\hat{S}_I\,,$$

with the orthogonal basis functions N_I, see Chapter 7.1 and Appendix 2- 4 for the similar Finite Element formulations.

So we can define a vector

$$F(\hat{S}) = (F_1(\hat{S}), F_2(\hat{S}), \cdots, F_N(\hat{S}))^\mathsf{T} \qquad \in \mathbb{R}^N \tag{6.21}$$

with the nodal unknowns

$$\hat{S} = (\hat{S}_1, \hat{S}_2, \cdots, \hat{S}_N)^\mathsf{T} \qquad \in \mathbb{R}^N \tag{6.22}$$

and the continuous functions

$$F_I(\hat{S}) = \langle h(|\nabla S_h^{n-1}|)\frac{S_h - S_h^{n-1}}{\tau}, N_I\rangle + \beta\langle D(S_h, S_h^{n-1}), N_I\rangle + \alpha\langle \nabla S_h, \nabla N_I\rangle \quad \text{for} \quad I = 1, N$$

to define the scalar product

$$F(\hat{S}^n)\cdot\hat{S}^{n-1} = \sum_{I=1}^{N} F_I(\hat{S}^n)\hat{S}_I^{n-1} = \begin{pmatrix} F_1(\hat{S}^n) \\ F_2(\hat{S}^n) \\ \vdots \\ F_N(\hat{S}^n) \end{pmatrix} \cdot \begin{pmatrix} \hat{S}_1^{n-1} \\ \hat{S}_2^{n-1} \\ \vdots \\ \hat{S}_N^{n-1} \end{pmatrix}. \tag{6.23}$$

If we can prove that $F(\hat{S})\cdot\hat{S} \geq 0$ for $|\hat{S}| = r$, this affirms the existence of a S_h^n in Eq. (6.15) as S_h^n is uniquely defined by $\hat{S}^n$.

For the timestep n and with the notation of Eq. (6.23) we express $F(\hat{S})\cdot\hat{S}$ by

$$\sum_{I=1}^{N} F_I(\hat{S})\hat{S}_I = \underbrace{\sum_{I=1}^{N} F_I(\hat{S})(\hat{S}_I - \hat{S}_I^{n-1})}_{(I)} + \underbrace{\sum_{I=1}^{N} F_I(\hat{S})\hat{S}_I^{n-1}}_{(II)}. \tag{6.24}$$

Now we argue that

$$\begin{aligned} (I) &= \langle h(|\nabla S_h^{n-1}|)\frac{S_h - S_h^{n-1}}{\tau}, S_h - S_h^{n-1}\rangle + \beta\langle D(S_h, S_h^{n-1}), S_h - S_h^{n-1}\rangle \\ &+ \alpha\langle \nabla S_h, \nabla(S_h - S_h^{n-1})\rangle \\ &\geq \frac{h}{\tau}||S_h - S_h^{n-1}||^2_{L^2(\Omega)} + \beta\int_\Omega \psi(S_h) - \psi(S_h^{n-1})\,dx + \frac{\alpha}{2}||\nabla S_h||^2_{L^2(\Omega)} \\ &- \frac{\alpha}{2}||\nabla S_h^{n-1}||^2_{L^2(\Omega)} + \frac{\alpha}{2}||\nabla(S_h - S_h^{n-1})||^2_{L^2(\Omega)}. \end{aligned}$$

For the last three terms, see Appendix 7.

Further we have that

$$(II) = \langle h(|\nabla S_h^{n-1}|)\frac{S_h - S_h^{n-1}}{\tau}, S_h^{n-1}\rangle + \beta\langle\frac{\psi(S_h) - \psi(S_h^{n-1})}{S_h - S_h^{n-1}}, S_h^{n-1}\rangle + \alpha\langle \nabla S_h, \nabla S_h^{n-1}\rangle. \tag{6.25}$$

With the weighted Young's inequality

$$ab \leq \rho a^2 + \frac{1}{4\rho} b^2 \tag{6.26}$$

and using Assumption 2, the first term on the right-hand side of Eq. (6.25) yields

$$\begin{aligned}&\langle S_h, \frac{h(|\nabla S_h^{n-1}|)}{\tau} S_h^{n-1}\rangle - \langle S_h^{n-1}, \frac{h(|\nabla S_h^{n-1}|)}{\tau} S_h^{n-1}\rangle \\ &\leq \rho ||S_h||_{L^2}^2 + \frac{\overline{h}}{4\rho\tau} ||S_h^{n-1}||_{L^2}^2 - \frac{\underline{h}}{\tau} ||S_h^{n-1}||_{L^2}^2 = \rho ||S_h||_{L^2}^2 + C_{\delta,1}(S_h^{n-1}) .\end{aligned}$$

The last term on the right-hand side of Eq. (6.25) can be estimated by

$$\alpha \langle \nabla S_h, \nabla S_h^{n-1}\rangle \leq \rho ||\nabla S_h||_{L^2}^2 + \frac{\alpha}{4\rho} ||\nabla S_h^{n-1}||_{L^2}^2 = \rho ||\nabla S_h||_{L^2}^2 + C_{\delta,2}(S_h^{n-1}) .$$

Now we need to look at the second term on the right-hand side of Eq. (6.25) containing the double well potential $\psi(S)$. From Assumption 3 we have that $|\psi''(S)| \leq C_{\psi,2}$ and thus we find a function with

$$\psi(S) \leq \frac{C_{\psi,2}}{2} S^2 + C_{\psi,3} , \quad C_{\psi,3} \in \mathbb{R} \tag{6.27}$$

and know that

$$\left| \frac{\psi(S_h) - \psi(S_h^{n-1})}{S_h - S_h^{n-1}} \right| \leq C_{\psi,4} |S_h| + C_{\psi,5} .$$

So we rewrite the second term on the right-hand side of Eq. (6.25) as

$$\begin{aligned}\beta \langle \left| \frac{\psi(S_h) - \psi(S_h^{n-1})}{S_h - S_h^{n-1}} \right|, |S_h^{n-1}|\rangle &\leq \beta \langle C_{\psi,4} |S_h| + C_{\psi,5}, |S_h^{n-1}|\rangle \\ &\leq \rho ||S_h||_{L^2}^2 + \frac{\beta C_{\psi,4}}{4\rho} ||S_h^{n-1}||_{L^2}^2 + \beta C_{\psi,5} ||S_h^{n-1}||_{L^2}^2 \\ &= \rho ||S_h||_{L^2}^2 + C_{\delta,3}(S_h^{n-1}) .\end{aligned}$$

Inserting these estimates in Eq. (6.25) yields

$$\text{(II)} \leq 2\rho (||S_h||_{L^2}^2 + ||\nabla S_h||_{L^2}^2) + C_{\delta,4}(S_h^{n-1}) . \tag{6.28}$$

Combining Eq. (6.28) with Eq. (6.25) we have

$$\begin{aligned}\text{(I)+(II)} \;\geq\; & \frac{\underline{h}}{\tau} ||S_h - S_h^{n-1}||_{L^2(\Omega)}^2 + \beta \int_\Omega \psi(S_h) - \psi(S_h^{n-1})\, dx + \frac{\alpha}{2} ||\nabla S_h||_{L^2(\Omega)}^2 \\ - \; & \frac{\alpha}{2} ||\nabla S_h^{n-1}||_{L^2(\Omega)}^2 + \frac{\alpha}{2} ||\nabla (S_h - S_h^{n-1})||_{L^2(\Omega)}^2\end{aligned}$$

$$
\begin{aligned}
&- \quad 2\rho(||S_h||_{L^2}^2 + ||\nabla S_h||_{L^2}^2) - C_{\delta,4}(S_h^{n-1}) \\
&\geq \quad \frac{h}{\tau}||S_h - S_h^{n-1}||_{L^2(\Omega)}^2 + \frac{\alpha}{2}||\nabla(S_h - S_h^{n-1})||_{L^2(\Omega)}^2 \\
&- \quad 2\rho(||S_h||_{L^2}^2 + ||\nabla S_h||_{L^2}^2) - C_{\delta,5}(S_h^{n-1}) \\
&\geq \quad \min\left\{\frac{h}{\tau}, \frac{\alpha}{2}\right\} ||S_h - S_h^{n-1}||_{H_1(\Omega)}^2 - \rho_2(||S_h||_{H_1(\Omega)}^2) - C_{\delta,5}(S_h^{n-1}) \\
&\geq \quad \tilde{C}_{min}(||S_h||_{H_1(\Omega)}^2 - ||S_h^{n-1}||_{H_1(\Omega)}^2) - \rho_2(||S_h||_{H_1(\Omega)}^2) - C_{\delta,5}(S_h^{n-1}) \\
&\geq \quad \left(\tilde{C}_{min} - \rho_2\right) ||S_h||_{H_1(\Omega)}^2 - C_{\delta,6}(S_h^{n-1}) \qquad (6.29)
\end{aligned}
$$

with $\tilde{C}_{min} = \tilde{C}C_{min}$, $\tilde{C} > 0$, $C_{min} = \min\left\{\frac{h}{\tau}, \frac{\alpha}{2}\right\}$.

In the penultimate step we used binomial formulas. Due to the fact that V_h is finite dimensional and S_h ist defined by $\hat{S}$, we can find an $r > 0$ such that for $\hat{S}$ with $r = |\hat{S}|$ it holds that $F(\hat{S}) \cdot \hat{S} > 0$. Due to the stated lemma "Zeros of a vector field" it follows that there exists a solution S_h^n fulfilling Eq. (6.15).

Additionally we have to show the uniqueness of a solution S_h^n. Let S_h^n and $\tilde{S}_h^n$ be two different solutions of Eq. (6.15). Using Assumption 2 and testing Eq. (6.15) with $v_{hS} = S_h^n - \tilde{S}_h^n$ we obtain

$$
\begin{aligned}
\frac{h(|\nabla S_h^{n-1}|)}{\tau}\langle S_h^n - \tilde{S}_h^n, S_h^n - \tilde{S}_h^n\rangle \quad &+ \quad \beta\langle D(S_h^n, S_h^{n-1}) - D(\tilde{S}_h^n, S_h^{n-1}), S_h^n - \tilde{S}_h^n\rangle \\
&+ \quad \alpha\langle \nabla S_h^n - \nabla\tilde{S}_h^n, \nabla S_h^n - \nabla\tilde{S}_h^n\rangle = 0\,. \qquad (6.30)
\end{aligned}
$$

With Assumption 3 and 2 and the Cauchy-Schwarz inequality we get

$$
\frac{h}{\tau}||S_h^n - \tilde{S}_h^n||_{L_2}^2 + \alpha||\nabla S_h^n - \nabla\tilde{S}_h^n||_{L_2}^2 \leq \beta||D(S_h^n, S_h^{n-1}) - D(\tilde{S}_h^n, S_h^{n-1})||_{L_2}||S_h^n - \tilde{S}_h^n||_{L_2}\,. \qquad (6.31)
$$

To estimate the term $||D(S_h^n, S_h^{n-1}) - D(\tilde{S}_h^n, S_h^{n-1})||_{L_2(\Omega)}^2$, we use the fundamental theorem of calculus

$$
\psi(S_h^n) - \psi(S_h^{n-1}) \leq \int_0^1 \psi'\left(S_h^{n-1} + \xi(S_h^n - S_h^{n-1})\right) \cdot (S_h^n - S_h^{n-1})\, d\xi
$$

leading to

$$
D(S_h^n, S_h^{n-1}) \leq \int_0^1 \psi'\left(S_h^{n-1} + \xi(S_h^n - S_h^{n-1})\right)\, d\xi\,.
$$

So we can calculate the difference

$$\begin{aligned}|D(S_h^n, S_h^{n-1}) \;-\; & D(\tilde{S}_h^n, S_h^{n-1})| \\ \leq\; & \int_0^1 \left|\psi'(S_h^{n-1} + \xi(S_h^n - S_h^{n-1})) - \psi'(S_h^{n-1} + \xi(\tilde{S}_h^n - S_h^{n-1}))\right| d\xi \\ \leq\; & \max\left\{|\psi''(S)| : S \in [a,b]\right\} |S_h^n - \tilde{S}_h^n| \\ \leq\; & \bar{d}|S_h^n - \tilde{S}_h^n| \,.\end{aligned}$$

For the last step we used Assumption 3 with $\bar{d} := C_{\psi,2}$.

So we conclude from Eq. (6.31) that

$$\underline{h}||S_h^n - \tilde{S}_h^n||^2_{L_2(\Omega)} + \alpha\tau||\nabla S_h^n - \nabla\tilde{S}_h^n||^2_{L_2(\Omega)} \leq \beta\bar{d}\tau||S_h^n - \tilde{S}_h^n||^2_{L_2(\Omega)} \,,$$

leading to

$$\left(\frac{\underline{h}}{\tau} - \beta\bar{d}\right)||S_h^n - \tilde{S}_h^n||^2_{L_2(\Omega)} + \alpha||\nabla S_h^n - \nabla\tilde{S}_h^n||^2_{L_2(\Omega)} \leq 0 \,.$$

For τ small enough, the left-hand side is positive and we must have $S_h^n = \tilde{S}_h^n$. Thus, the lemma is proven. □

7 Numerical validation

With Eq. (6.15) we have a fully discretised variational formulation of the phase field equation, which we can implement numerically. We will explain the implementation in Chapter 7.1, using two-dimensional Finite Elements.

In this section, we will look at some simple one-dimensional tests that verify basic features of the models. This one-dimensional implementation will not be explained in detail.

First, we set $W_S = 0$, $\mu = \lambda = \nu = 1$ and choose the mobility functions f_1 and f_2 as explained in Table 5.2 with $c = 1$. Accordingly, the phase field formulations Eq. (5.5) and Eq. (5.6) read

$$\partial_t S \;=\; -c_1\,(\psi'(S) - \Delta S), \tag{7.1}$$

$$\partial_t S \;=\; -(\psi'(S) - \Delta S)|\nabla S| \,, \tag{7.2}$$

with $c_1 = 0.4714$ calculated by Eq. (2.17).

Although the Allen-Cahn model and the hybrid model are seen as special cases of the general model in Chapter 5, we want to go back one step and consider the models individually.

We drop the separation term, setting $\psi(S) = 0$, and investigate the equations

$$\partial_t S = c_1 \Delta S, \tag{7.3}$$

$$\partial_t S = \Delta S |\nabla S|. \tag{7.4}$$

We recognise in Eq. (7.3) the structure of the homogeneous heat conduction equation, a partial differential equation of parabolic type. For the one-dimensional case it is given by the equation

$$\partial_t u_{heat}(x,t) - \alpha_{heat} \partial_{xx} u_{heat}(x,t) = 0\,, \tag{7.5}$$

with a constant thermal diffusion coefficient α_{heat} and $u_{heat}(x,t)$ for the unknown temperature.

Fig. 7.1 shows the stationary solution of the numerical simulation of Eq. (7.3) and Eq. (7.4), completed by the initial- and Dirichlet boundary conditions

$$\begin{aligned} S(0,x) &= \text{Heaviside}(x)\,, & (7.6)\\ S(t,-3) &= 0\,, & (7.7)\\ S(t,3) &= 1\,. \end{aligned}$$

Both plots (coloured red and blue) lie exactly on each other. We see that the numerical solutions of Eq. (7.3) and Eq. (7.4) tend to the stationary solution of the one-dimensional heat equation. The development of the solution of the hybrid model is based on the jump at $x = 0$ of the initial condition and its numerical implementation in a finite dimensional grid. As we have explained, a gradient term $\nabla S = 0$ in fact prevents an evolution of the order parameter S in time.

Next, we switch on the double well function Eq. (2.5) as separation term and study numerical solutions of the Allen-Cahn equation (7.1) and the hybrid equation (7.2) with the initial and boundary conditions Eq. (7.6). We refer to Eq. (5.29) and calculate for $\mu = \lambda = \nu = 1$ the width relation

$$re_A = \frac{\text{width}_{AC}}{\text{width}_H} = \frac{(\mu\lambda)^{1/2}}{\nu^{1/2}} = 1\,. \tag{7.8}$$

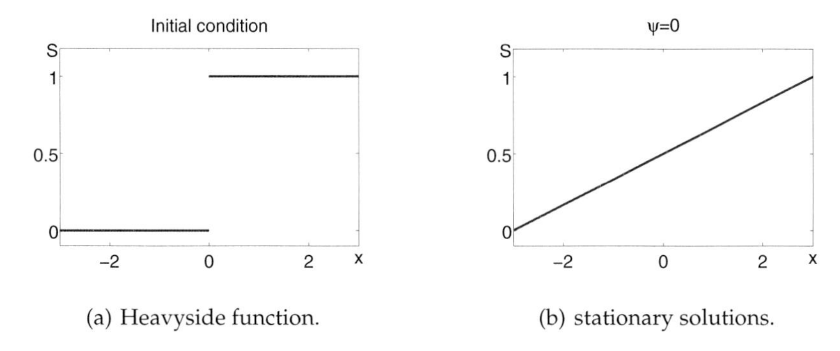

(a) Heavyside function. (b) stationary solutions.

Figure 7.1: Initial condition and stationary solution of the hybrid model and the Allen-Cahn model without double well potential.

The relation is in agreement with the fact that the numerically simulated curves in Fig. 7.2 lie exactly on each other. The length of the interval of x belonging to $S \in]0,1[$ represents the width of the interface and is exacty the same for both models. As explained before, the parameters ν and μ determine the width of the interface that has to be resolved numerically.

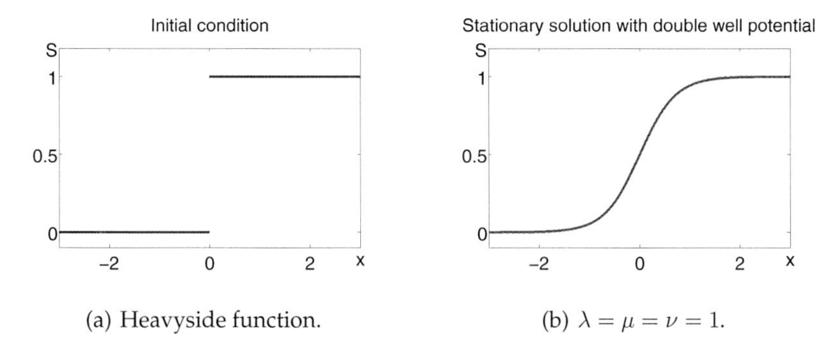

(a) Heavyside function. (b) $\lambda = \mu = \nu = 1$.

Figure 7.2: Initial condition and stationary solution of the hybrid model and the Allen-Cahn model with double well potential.

To improve the solution and to check the condition Eq. (5.29) for the width of the transition zone between the phases, we set $\lambda = \mu = \nu = 0.1$ and calculate $r_{eA} = \frac{\text{width}_{AC}}{\text{width}_H} = \frac{(\mu\lambda)^{1/2}}{\nu^{1/2}} \approx 0.316$. Fig. 7.3 confirms this result.

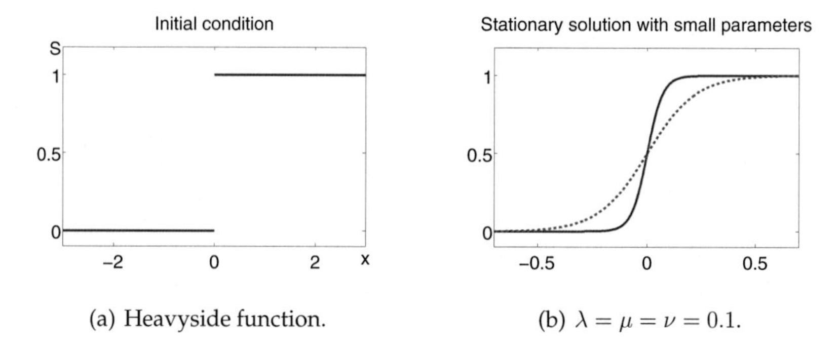

(a) Heavyside function.

(b) $\lambda = \mu = \nu = 0.1$.

Figure 7.3: Initial condition and stationary solution of the hybrid model (red dotted line) and the Allen-Cahn model (blue solid line).

The dotted red line belongs to the hybrid model which shows an about three times wider transition zone. Both curves meet at the point $(0, 0.5)$, denoting the midpoint of the transition zone between the two phases given by Eq. (1.2).

This simple numerical study confirms that choosing the model parameters very small, the mesh can be adjusted coarser for the hybrid model.

Another interesting point is that the solution for a constant initial condition $S(0, x) = c$ and the Dirichlet boundary conditions $S(t, 3) = S(t, -3) = c$ differs for both models, see Fig. 7.4. Only for the initial condition $S(0, x) = 0.5$ the values of

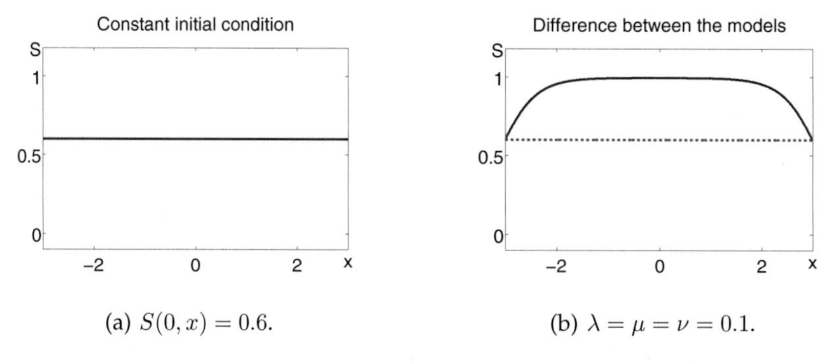

(a) $S(0, x) = 0.6$.

(b) $\lambda = \mu = \nu = 0.1$.

Figure 7.4: Initial condition and stationary solution of the hybrid model (red dotted line) and the Allen-Cahn model (blue solid line).

the Allen-Cahn model remain constant (but might change as soon as we apply a small singular disturbance as first test simulations showed).

This confirms that an interface simulated by the hybrid model does not move as long as $\nabla S = 0$ while the Allen-Cahn model (without gradient term in the mobility function) forces the order parameter into the value $S(t,x) = 1$ for initial values $S(0,x) > 0.5$ and into the value $S(t,x) = 0$ for initial values $S(0,x) < 0.5$.

To complete our one-dimensional studies, we verify that the free energy decreases in time. To prove this, we track solutions of the Allen-Cahn model and the hybrid model comparing their energies. The results are shown in Table 7.4.

Since the values $\nu = \mu = \lambda = 1$ are chosen large, they cause a large error for the hybrid model compared to the Allen-Cahn model, so its curve descends slower. For the same reason the Allen-Cahn model forms a smoother curve (blue solid line) while the curve of hybrid model (red dotted line) remains more angular.

Calculating the energy values by Eq. (5.1), we see that they decay as we expected.

Table 7.4: Numerical evidence of the energy decay property for the Allen-Cahn model (blue solid line) and the hybrid model (red dotted line).

$t = 0$	$t = 0.1$	$t = 1$	$t = 5$	$t = 30$
$\bar{E}(S)_{AC}/$[units]	2.19	0.93	0.05	0
$\bar{E}(S)_{H}/$[units]	4.49	1.00	0.97	0
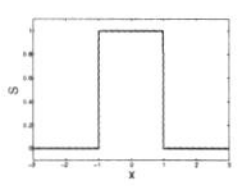	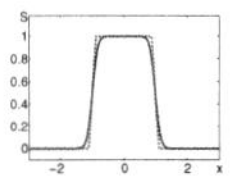	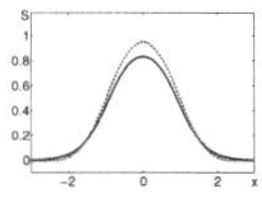	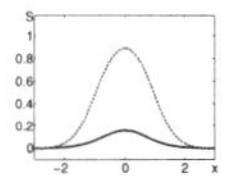	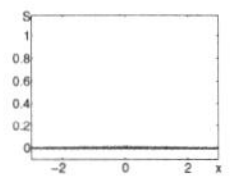
$\lambda = \mu = \nu = 1,\ dt = 0.01,\ dx = 0.01$				

Zooming into the point $(0,0)$ in the last diagram, we recognise that the value of S at $x = 0$ does not reach 0 for the hybrid model. Further calculations with finer grids cannot completely eliminate this fact, see Fig. 7.5. We will come back to this point in Chapter 7.2.

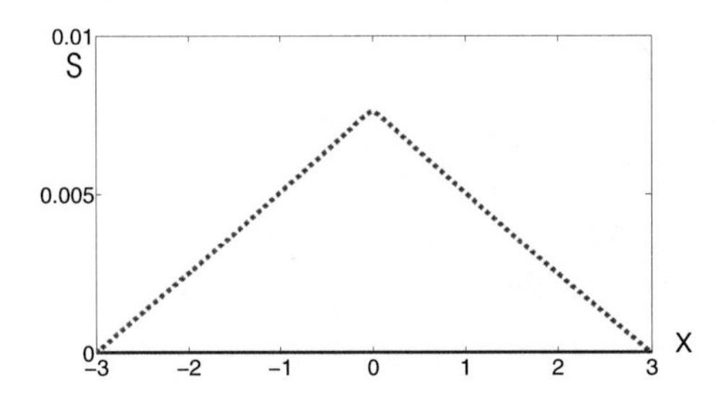

Figure 7.5: Critical size at $(0, 0)$ - hybrid model (red dotted line) and Allen-Cahn model (solid blue line) using very fine discretisation in x and t.

7.1 Two-dimensional numerical implementation with the Finite Element method

Now we introduce the two-dimensional implementation of the general model by the Finite Element method.

In literature there is a huge amount of publications on numerical schemes for Allen-Cahn type models. Implicit schemes for two- and three-dimensional numerical formulations proving energy properties for the Allen-Cahn and the Cahn-Hilliard equation can be found, e.g. in [41]. More implicit and explicit schemes concerning the implementation of the Allen-Cahn model are given in [58], [86].

We also will use an implicit scheme to solve a non-linear system applying a Banach fixed-point iteration. Therefore we rewrite Eq. (6.15) as fixed-point equation

$$\langle S_h^n, v_{hS} \rangle = \underbrace{-\frac{\beta}{h(|\nabla S_h^{n-1}|)} \tau \langle D(S_h^n, S_h^{n-1}), v_{hS} \rangle - \frac{\alpha}{h(|\nabla S_h^{n-1}|)} \tau \langle \nabla S_h^n, \nabla v_{hS} \rangle + \langle S_h^{n-1}, v_{hS} \rangle}_{\Phi(\langle S_h^n, v_{hS} \rangle)} \tag{7.9}$$

to implement it for numerical considerations.

Using the Finite Element method, explained in more detail in Appendix 2-4, we define approximate solutions for the respective time step (index n) by

$$S_h(x) = \sum_{I=1}^{N} N_I(x)\hat{S}_I \tag{7.10}$$

with N for the number of global nodes. The "hat"-Symbol $(\hat{\cdot})$ stands for nodal values of the unknowns and the index $(\cdot)_h$ denotes the interpolated continuous function. The unknowns $\hat{S}_I$ are the coefficients multiplying the shape functions $N_I(x)$ and the test functions are along the lines of Eq. (7.10) given by

$$v_{hS}(x) = \sum_{I=1}^{N} N_I(x)v_{SI}\,. \tag{7.11}$$

The shape functions N_I have values 1 or 0 at the respective node I and we have $S_h(I) = \hat{S}_I$. The local shape functions refer to two-dimensional four-node element we implemented in MATLAB, see Appendix 2.

We go back one step and insert Eq. (7.10) and Eq. (7.11) into the semi-discrete variational formulation of the general model Eq. (6.1) leading to

$$M_{2_{mod}}\hat{S}_t + \alpha K\hat{S} = -\beta M_1 F(\hat{S}) \tag{7.12}$$

with mod standing for the respective model and $\hat{S}_t \in \mathbb{R}^N$ containing the time derivatives of the nodal order parameters. $M_{2_{mod}}$ and M_1 stand for the (weighted) mass matrices and K for the stiffness matrix, see Appendix 2 - 4.

Inserting the respective matrix $M_{2_{mod}}$ yields

$$M_{2_{AC}}\hat{S}_t + \alpha K\hat{S} = -\beta M_1 F(\hat{S}) \tag{7.13}$$

for the Allen-Cahn model and

$$M_{2_H}\hat{S}_t + \alpha K\hat{S} = -\beta M_1 F(\hat{S})\,. \tag{7.14}$$

for the hybrid model.

For further details on the Finite Element method, see Appendix 2 - 4, [23], [88] and [15].

The extension to the fully-discrete formulation of Eq. (7.12) is only one step and will be shown below, but first we consider the term $F(\hat{S})$ on the right-hand side of Eq. (7.13) and Eq. (7.14).

The vector F contains the respective difference quotients of the double well potential, defined by Eq. (6.13). We implement the difference quotient instead of deriving ψ, defined by Eq. (2.5), in order to be concordant with our theoretical discussion on the well-posedness and the energy decay properties in the previous chapters.

We calculate

$$\begin{aligned}
\frac{\psi(S_h^n)-\psi(S_h^{n-1})}{S_h^n-S_h^{n-1}} &= \frac{4\left((S_h^n)^4-2(S_h^n)^3+(S_h^n)^2\right)-4\left((S_h^{n-1})^4-2(S_h^{n-1})^3+(S_h^{n-1})^2\right)}{S_h^n-S_h^{n-1}} \\
&= \frac{4\left(\left((S_h^n)^4-(S_h^{n-1})^4\right)-2\left((S_h^n)^3-(S_h^{n-1})^3\right)+\left((S_h^n)^2-(S_h^{n-1})^2\right)\right)}{S_h^n-S_h^{n-1}} \\
&= 4\left((S_h^n)^2+(S_h^{n-1})^2\right)\left(S_h^n+S_h^{n-1}\right)-\frac{8((S_h^n)^3-(S_h^{n-1})^3)}{S_h^n-S_h^{n-1}}+4(S_h^n+S_h^{n-1}) \\
&= 4\left((S_h^n)^2+(S_h^{n-1})^2+1\right)\left(S_h^n+S_h^{n-1}\right)-8\left((S_h^n)^2+S_h^nS_h^{n-1}+\left(S_h^{n-1}\right)^2\right),
\end{aligned}$$

so we have

$$\begin{aligned}
\frac{\psi(S_h^n)-\psi(S_h^{n-1})}{S_h^n-S_h^{n-1}} &= 4\left((S_h^n)^3+d_1\,(S_h^n)^2+d_2\,S_h^n+d_3\right) \qquad (7.15)\\
\text{with}\quad d_1 &= -2+S_h^{n-1}, \\
d_2 &= 1-2S_h^{n-1}+\left(S_h^{n-1}\right)^2, \\
d_3 &= S_h^{n-1}-2\left(S_h^{n-1}\right)^2+\left(S_h^{n-1}\right)^3.
\end{aligned}$$

Now we can discretise Eq. (7.12) in time. For the fully-discrete scheme we approximate the time derivative of the nodal order parameter by

$$\hat{S}_{t_I}=\frac{\hat{S}_I^n-\hat{S}_I^{n-1}}{\tau}. \qquad (7.16)$$

Inserting Eq. (7.16) in Eq. (7.12) yields

$$M_{2_{mod}}\frac{\hat{S}^n-\hat{S}^{n-1}}{\tau}+\alpha K\hat{S}^n=-\beta M_1F(\hat{S}^n) \qquad (7.17)$$

with the index $mod = AC$ for the Allen-Cahn model and $mod = H$ for the hybrid model. Eq. (7.17) can be transformed to

$$\hat{S}^n \underbrace{\left(M_{2_{mod}} + \alpha\tau K\right)}_{K_{mod}} = M_{2_{mod}}\hat{S}^{n-1} - \beta\tau M_1\, F(\hat{S}^n) \tag{7.18}$$

and for the positive definite matrices $M_{2_{mod}}$ and K we can calculate

$$\hat{S}^n = K_{mod}^{-1}\left(M_{2_{mod}}\hat{S}^{n-1} - \beta\tau M_1\, F(\hat{S}^n)\right) . \tag{7.19}$$

This scheme is explained in Table 7.5. We implement the two to be compared phase field models with $\alpha = \mu^{1/2}\lambda$ and $\beta = \mu^{-1/2}$ for the Allen-Cahn model and $\alpha = \nu$ and $\beta = 1$ for the hybrid model.

Table 7.5: Implicit scheme.

Initialisation: $\hat{S}^1 = S(0,x)$, tol=1e-2, err=1		
Loop n=1,Nt		
	Calculate K, M_1	
	k=-1, $\hat{S}_{k+1} = \hat{S}^n$	
		While err $>$ tol
		$k = k + 1$
		Solve $\hat{S}_{k+1} = K_{mod}^{-1}(\hat{S}^n)\left(M_{2_{mod}}(\hat{S}^n)\hat{S}^n - \beta\tau M_1\, F(\hat{S}_k, \hat{S}^n)\right)$
		err $= \dfrac{\lvert\lvert\hat{S}_{k+1} - \hat{S}_k\rvert\rvert \max}{\lvert\lvert\hat{S}_{k+1}\rvert\rvert \max}$
		end
	$\hat{S}^{n+1} = \hat{S}_{k+1}$	
end		

7.2 Shrinking circle and comparison to an analytical solution

As first numerical example we use the analytical solution of a shrinking circle derived in [4]. Inside the circle, we define the first phase by the order parameter $S(x,t) = 1$ and the phase outside the circle has the value $S(x,t) = 0$. Omitting inner and outer forces and the coupling to some constitutive material equation, the shrinking is due only to the curvature term.

We recall Eq. (5.23) and Eq. (5.24) and compare both models in terms of the curvature dependent driving force, setting $\nu = \lambda$. The definition of ω_1 is given in Eq. (3.18), based on the considerations in [4], using a slightly different double well potential $\tilde{\psi}_1$. As explained in Chapter 5.3 we set ω_1 equal to c_1.

The formulation (6.15) is implemented as explained in Chapter 7.1 with the respective parameters for each model. For the boundary condition we set

$$S(x,t) = 0 \quad \text{for} \qquad x \in \partial\Omega \tag{7.20}$$

and the initial conditions are given as

$$S(x,0) = S_0 \qquad \text{with} \quad S_0(x,y) = \max\{0; 3 - 2.5((x-1.5)^2 + (y-1.5)^2)\} \tag{7.21}$$

inside the domain

$$\bar{\Omega} = \{(x,y) : 0 \leq x \leq 3 \,, \, 0 \leq y \leq 3\}.$$

Using Eq. (5.23) and Eq. (5.24) for an analytic solution, we drop the error terms and inserting $s_{1_{err}} = s_{2_{err}} = -r_t$ for the time derivative of the radius and $\kappa_\Gamma = 1/r$ for the curvature yields

$$r_t = -p\,\frac{1}{r} \qquad \text{with} \quad p = \lambda^{1/2} c_1 = \nu^{1/2} \omega_1 \,. \tag{7.22}$$

With the initial condition

$$r(t=0) = 1 \,, \tag{7.23}$$

the analytic solution of this first order ordinary differential equation follows by

$$r(t) = \sqrt{1 - 2\,p\,t} \,. \tag{7.24}$$

Calculating discrete solutions of Eq. (7.19) and tracking the radius evolution over time, gives us the interface velocities, expressed by Eq. (5.23) and Eq. (5.24), including the error terms $O(\mu^{1/2})$ and $O(\nu)$ for the centerline of the interface given by Eq. (1.2) (with the value $S = 0.5$). We compare these numerical solutions to the analytical solution (7.24). This procedure is illustrated in Fig. 7.6 (a)-(d), where we see the circle with the inner phase with $S = 1$ (red) and the outer phase with $S = 0$ (blue) and the circular diffuse interphase. We measure the radius r as the distance from the midpoint to the contour line of $S = 0.5$. The value of S changes smoothly within the interface, even though the plots show contour lines (due to technical reasons). The difference between the analytical and numerical solution, which can be seen in Fig. 7.6 (d), is due to error terms from Eq. (5.25) and Eq. (5.26), as well as other numerical errors, such as discretisation errors, round-off errors, truncation errors and approximation errors. Since we can keep these other errors comparatively small through choice of our elements, modelling and refinement, we neglect them and concentrate on the error terms explained in chapter 5.4 and their influence on the accuracy of the solution (depending on the spatial and temporal discretisation).

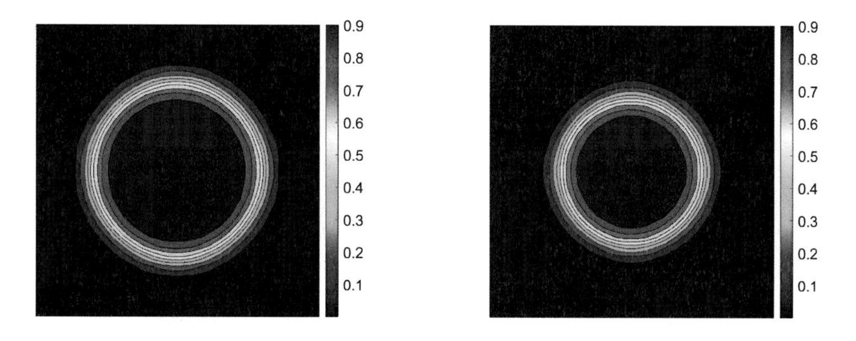

(a) order parameter S with $r(t_1 = 0.2) \approx 0.9$ (b) order parameter S with $r(t_2 = 0.5) \approx 0.76$

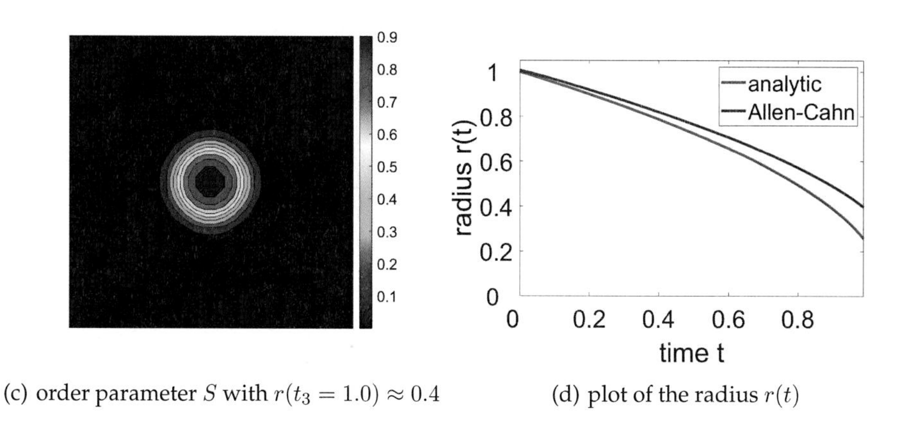

(c) order parameter S with $r(t_3 = 1.0) \approx 0.4$ (d) plot of the radius $r(t)$

Figure 7.6: Comparison of the analytical and the numerical solution.

In Fig. 7.7, the numerical and the analytical solutions of the shrinking circle for the Allen-Cahn model and the hybrid model are compared for the given time step $dt = 0.001$, the parameters $\mu = \nu = \lambda = 0.1$ and different spatial discretisations. We specify the mesh size $dx = a/nx$ with nx for the number of nodes per direction and $a = 3.0$ for the width of the domain.

We used the implicit scheme from Table 7.5.

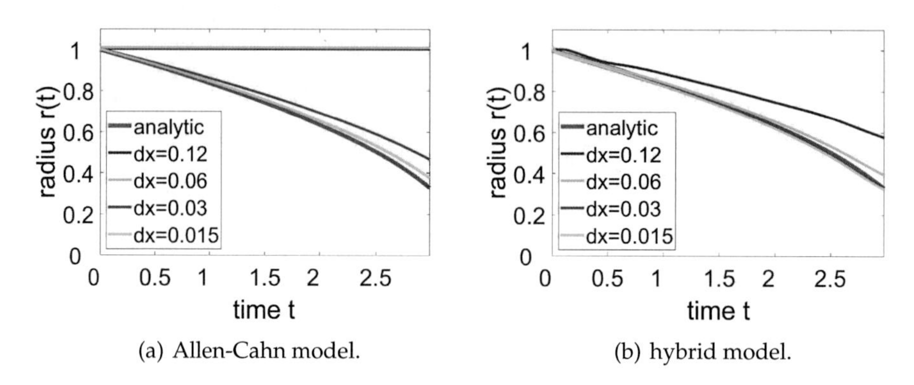

(a) Allen-Cahn model.

(b) hybrid model.

Figure 7.7: Convergence study with $dt = 0.001$, $\lambda = \mu = \nu = 0.1$.

Fig. 7.7 shows that the hybrid model converges faster on the same grid compared to the Allen-Cahn model for the given rather small interfacial energy. This can be understood by the fact that the numerical interface width of the hybrid model is wider and therefore, the interface area is numerically better resolved.

Eq. (5.25) and Eq. (5.26) predict the same error order resulting from the model- and the residual error by choosing $\lambda = \mu = \nu$. Eq. (5.30) predicts that the widths of the interfaces, we indicate by the letter d from now on, behave like $re_B = \nu^{1/2}$, see Eq. (5.30). We obtain $re_B = 0.316$ choosing $\nu = 0.1$, which explains the wider transition zone for the hybrid model.

Since the time step was chosen sufficiently small, the models should converge to the same values by refining the meshes accordingly. Regarding Eq. (5.30) we should achieve the same convergence behaviour by choosing the mesh three times coarser for the hybrid model.

Fig. 7.8 shows simulation results using the spacial discretisation, defined by $nx = 120$ for the Allen-Cahn model and $nx = 40$ for the hybrid model and their widths d at $t = 0.1$. Apparently, the relation of their widths d is of order $\approx 1:3$, whereas their solutions, seen in Fig. 7.8(c), are quite similar and shrink with a comparable velocity.

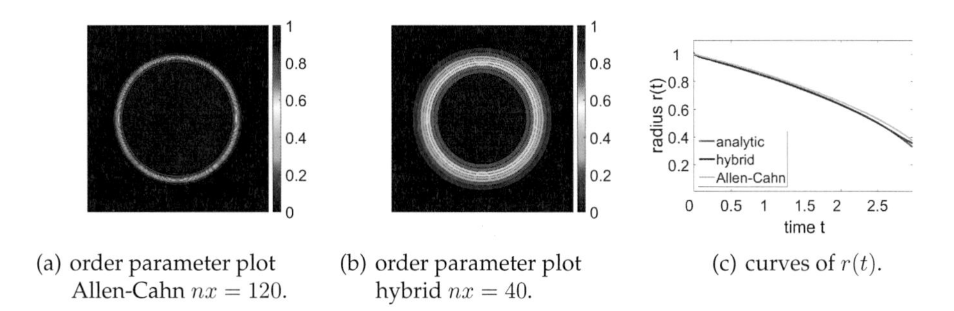

(a) order parameter plot Allen-Cahn $nx = 120$.

(b) order parameter plot hybrid $nx = 40$.

(c) curves of $r(t)$.

Figure 7.8: Comparison with $\lambda = \mu = \nu = 0.1$.

To adjust the width d of the hybrid model to the width d of the Allen-Cahn model, Eq. (5.29) predicts the width relation

$$1 = \frac{(\mu\lambda)^{1/2}}{\nu^{1/2}}. \tag{7.25}$$

Thus, we have to choose $\nu = 0.01$, keeping $\mu = \lambda = 0.1$. The widths d compared for this case are shown in Fig. 7.9.

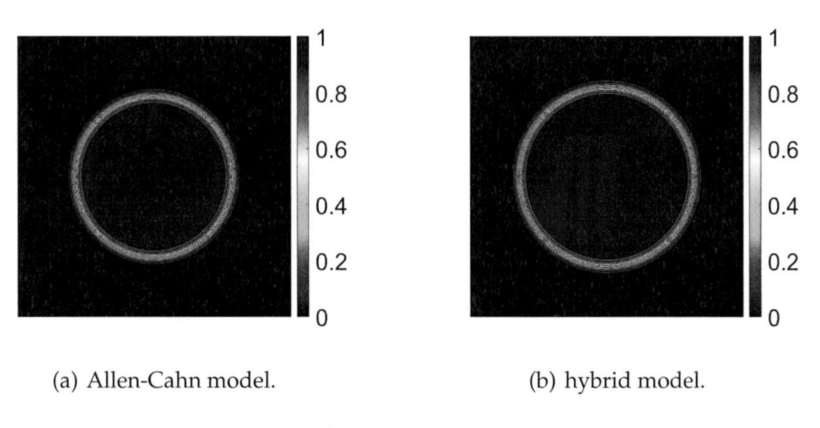

(a) Allen-Cahn model.

(b) hybrid model.

Figure 7.9: Width of interface d for $re_A = 1$ at $t = 1$.

Note that the corresponding analytical solutions differ due to different values for ν and μ. The numerical solutions in Fig. 7.10 both follow the analytical solutions quite well.

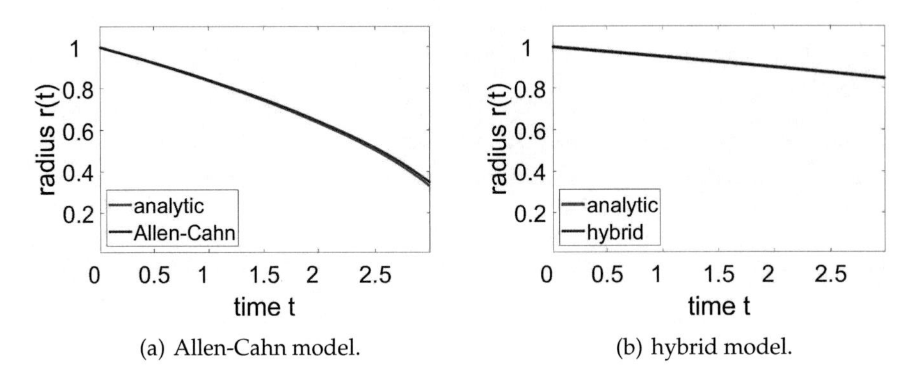

(a) Allen-Cahn model.

(b) hybrid model.

Figure 7.10: $\nu = 0.01, \mu = \lambda = 0.1 \rightarrow re_A = 1, dt = 0.001, dx = 0.01.$

Finally, the theory is confirmed by the given examples. The widths d, shown in the respective solution plots, depend on the parameter μ and ν. For small values of λ, defining the interfacial energy, the hybrid model has a numerical advantage as explained in Chapter 5.4. In this case, we can choose a coarser mesh for the hybrid model, whereas for the Allen-Cahn model a finer mesh is needed for comparable results.

In case of high interfacial energy values we have to be careful using the hybrid model. The error analysis in Chapter 5.4 explained the dependency of the error terms on the interfacial energy density. This has to be considered by choosing the best model.

Critical size. Concluding, we come to one point which we already identified as a possible difficulty for the hybrid model in the beginning of this chapter, see Fig. 7.5. If we let the radius go to zero, the hybrid model behaves somehow strange. Simulating a completely shrinking, thus dissolving circle, the numerical hybrid model curve for the radius branches off closely before the radius becomes zero, see Fig. 7.11.

After discussing with the author of [44], we tried a simulation with a new rescaling for the hybrid model, namely

$$\partial_t S = -c\frac{1}{\varepsilon_\nu}(\psi'(S) - \nu\Delta S)|\nabla S|\,, \quad \nu = \varepsilon_\nu^2\,. \tag{7.26}$$

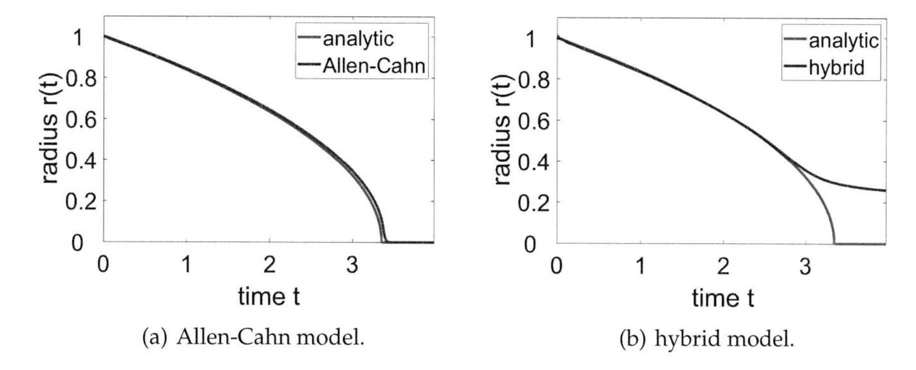

(a) Allen-Cahn model.

(b) hybrid model.

Figure 7.11: Comparison with $dt = 0.001, dx = 0.01$, $\nu = \mu = \lambda = 0.1$.

Following the asymptotic solution analysis in [34], we obtain an expression of the normal velocity of the diffuse interface of the hybrid model in terms of

$$s_{h_G} \approx c\omega_1 \kappa_\Gamma \,. \tag{7.27}$$

The advantage of this rescaling is that the normal interface velocity gets rid of the asymptotic parameter ν in Eq. (5.24) that was originally introduced as an asymptotic solution parameter and not as a interfacial energy parameter. The idea was that sending $\nu \to 0$ stopped the shrinking at a certain time point and that a formulation without ν in the analytical solution of s_{h_G} could adjust this solution to the solution of the Allen-Cahn model.

The implementation of Eq. (7.26) and Eq. (7.27) and the comparison with the former calculation of the hybrid model of [4] unfortunately did not solve the problem of the non-disappearing circle, see Fig. 7.12.

To understand which parameters influence the point in time of the almost stagnation t_{st}, we varied the parameter ν determining the interface width d. We recognice a large influence of ν on the point in time t_{st} related to the corresponding interface width d. The adjustment to a smaller numerical interface width, done by the choice of a smaller ν, reduces the radius $r_{st} \approx d$, belonging to the time t_{st}, see Fig. 7.13. Refining the mesh discretisation, however, has almost no influence on this fact, see Fig. 7.14.

So it seems as if the circle stops shrinking as soon as the inner phase, indicated by $S = 1$, disappears and accordingly, the values of the radii of the hybrid model, shown in Figs. 7.12 – 7.14, converge to the width d. The fact that the radii

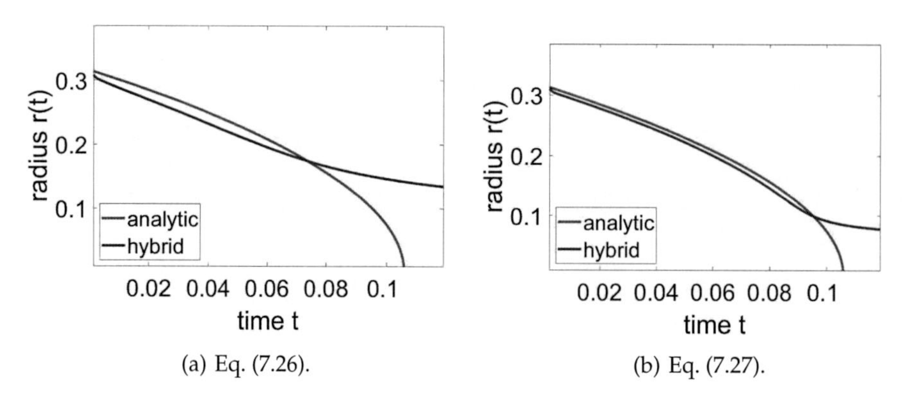

(a) Eq. (7.26).

(b) Eq. (7.27).

Figure 7.12: hybrid model with Eq. (7.27) and $dt = 0.001$, $dx = 0.01$, $\nu = 0.01$.

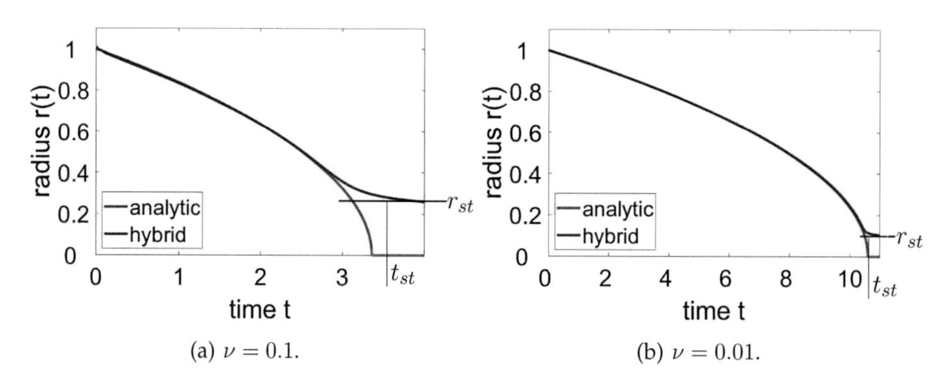

(a) $\nu = 0.1$.

(b) $\nu = 0.01$.

Figure 7.13: Hybrid model $dt = 0.001, dx = 0.01$.

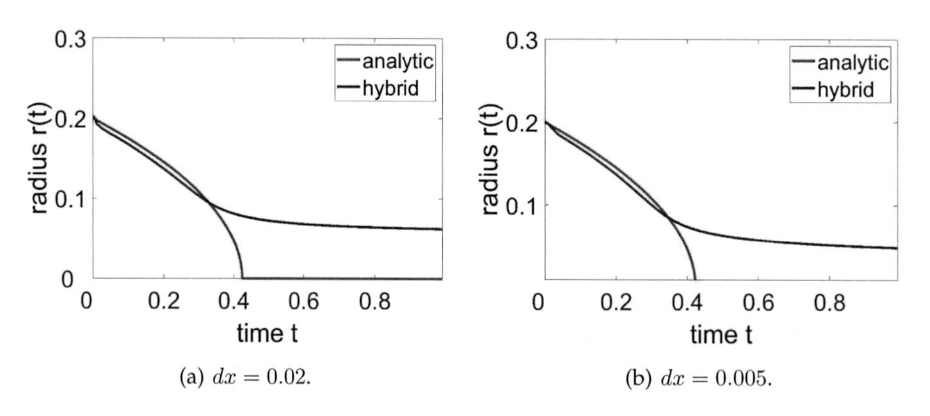

(a) $dx = 0.02$.

(b) $dx = 0.005$.

Figure 7.14: Hybrid model $dt = 0.001, \nu = 0.01$.

converge to zero after a long time (not shown here) was not further examined. We can only guess that this is due to numerical effects.

At this point, we state that the hybrid model shows an odd behaviour on small scales and we will explore this point further in the last chapter of Part III. With these final conclusions and open questions we close Part II and turn to the elastic phase field formulation introduced in Part III.

PART III: SYSTEMS COUPLED TO ELASTICITY

8 Elastic phase field model

In this section, we leave the solely curvature driven phase field situation and extend the models to linear elasticity. The effects of inner and outer forces on the interface development are overlapping with the influence of interface properties like interface energy and curvature. In some cases the interface terms can be neglected compared to the terms of the inner and outer forces, in other situation they are quite important, see [67], [42].

In the introduction in Chapter 1 we gave some examples of inner and outer forces and related problems. A further application of coupled phase field research is the topology optimisation under mass conservation, see [17] and another use is the development of nickel-base superalloys and their behaviour under pressure and thermal loads, see [71]. Besides, materials scientists are interested in the development of crystal twinning under mechanical loadings, see [56], to give just a few examples of the countless applications.

The coupling to constitutive material laws can be extended to plasticity, viscoelasticity, piezoelectricity and others. For simplicity, we restrict our considerations to linear elasticity.

Linear elasticity An elastic body undergoes shape changes due to inner strains or outer forces. In the theory of linear elasticity, the original shape of the deformed material is regained as soon as the applied forces are removed (reversible deformation). In materials science, rheological models explain the restoring forces between the individual material particles by an idea of elastic springs.

Furthermore, the theory of plastic deformations describes the situation where defects migrate and permanent deformations remain above a material dependent yield stress (irreversible deformation). Such plasticity effects are not examined in the present work, though they could also be coupled to the phase field models by constitutive equations.

To extend the general phase field equation (5.2) to an elastic formulation we need an additional elastic energy term $W(S, \varepsilon(u))$. To emphasise that the strain ε depends on the displacement u, we will use the notation $\varepsilon(u)$ from now on. With the notations of Chapter 5, knowing that $S = S(x,t)$ and $u = u(x,t)$, the global free energy for an elastic phase field formulation can be defined by

$$\bar{E}_{el}(S, \varepsilon(u)) = \int_{\Omega} W(S, \varepsilon(u)) + \beta\psi(S) + \frac{\alpha}{2}|\nabla S|^2 \, dx \,, \tag{8.1}$$

with the respective parameters α and β. The elastic phase field equation

$$h(|\nabla S|)\partial_t S = -W_S(S, \varepsilon(u)) - \beta\psi'(S) + \alpha\Delta S \tag{8.2}$$

arises as gradient flow of the free energy (8.1) as explained in Chapter 2. The elastic energy $W(S, \varepsilon(u))$ was defined by Eq. (2.3) and we denoted its partial derivative with respect to S by W_S. Its derivative with respect to the strain ε led to the definition of the Cauchy stress tensor, introduced by Eq. (2.20). We now extend this stress tensor to a dependency of S in terms of

$$T(S, \varepsilon(u)) = \mathbb{C}(S)(\varepsilon(u) - \bar{\varepsilon}S) \qquad \in \mathcal{S}^3 \,. \tag{8.3}$$

The included phase dependent strain tensor is given by

$$\tilde{\varepsilon}(S, u) = \varepsilon(u) - \bar{\varepsilon}S \qquad \in \mathcal{S}^3 \,, \tag{8.4}$$

as a function of the unknowns $u \in \mathbb{R}^3$, see Eq. (2.4), and $S \in \mathbb{R}$. Eq. (8.3) is also known as Hooke's law, see [50] for more details.

The constant tensor $\bar{\varepsilon} \in \mathcal{S}^3$ is used in physical applications, for example within the description of metal alloys, where each phase contains an own eigenstrain due to misfitting atomic lattices. In the present work, the phase with $S = 0$ has a zero eigenstrain and the phase with $S = 1$ has a constant eigenstrain $\bar{\varepsilon}$, being part of the strain $\tilde{\varepsilon}(S, u)$, defined by Eq. (8.4).

In parallel, we construct a dependency of the elasticity tensor $\mathbb{C}$ on the order parameter S to define the elastic properties as a function of the respective phase in terms of

$$\mathbb{C}(S) = \mathbb{C}_1 + S(\mathbb{C}_2 - \mathbb{C}_1) \,. \tag{8.5}$$

Primary, $\mathbb{C}$ is a fourth order tensor, but for the implementation of the model we reduce $\mathbb{C} : \mathcal{S}^3 \to \mathcal{S}^3$ to a tensor of second order, see Appendix 5 and [53]. We assume constant entries in the elasticity tensor $\mathbb{C}_1$ for phase 1 and $\mathbb{C}_2$ for phase 2 and preserve a smooth transition represented by the values of $\mathbb{C}(S)$ within the diffuse interface. We assume the extension of $\mathbb{C}(S)$ outside of $I = [0,1]$ to be linear, symmetric and positive definite.

With these definitions, the elastic energy for an elastic phase field problem is given by

$$W(S,\varepsilon(u)) = \frac{1}{2}(\varepsilon(u) - \bar{\varepsilon}S)^{\mathsf{T}} : ((\mathbb{C}_1 + S(\mathbb{C}_2 - \mathbb{C}_1))(\varepsilon(u) - \bar{\varepsilon}S)) \,. \tag{8.6}$$

In Part II we studied the phase field models containing as unknown only the order parameter S. To extend this theory to a coupled material problem of linear elasticity we need an additional constitutive material law and the corresponding balance equation. For the equilibrium condition we have

$$-\nabla \cdot T(S,\varepsilon(u)) = b \tag{8.7}$$

with $b = b(x)$ as a given time-independent force field on the right-hand side.

Adopting the initial- and boundary conditions Eq. (2.15), we need additional boundary conditions for the additional unknowns u. For our later numerical simulations, we choose Dirichlet boundary conditions

$$u(x) \;=\; u_D(x)\,, \qquad x \in \partial\Omega\,, \quad \forall t. \tag{8.8}$$

Thus, we state the following partial differential equation system

$$\begin{aligned}
-\nabla \cdot T(S,\varepsilon(u)) &= b\,, && (8.9)\\
T(S,\varepsilon(u)) &= \mathbb{C}(S)(\varepsilon(u) - \bar{\varepsilon}S)\,, && (8.10)\\
h(|\nabla S|)\partial_t S &= -W_S(S,\varepsilon(u)) - \beta\psi'(S) + \alpha\Delta S\,, && (8.11)\\
S(x,t) &= 0\,, \quad x \in \partial\Omega\,, \quad t \geq 0\,, && (8.12)\\
S(x,0) &= S_0\,, \quad x \in \Omega\,, && (8.13)\\
u(x) &= u_D(x)\,, \quad x \in \partial\Omega\,, \quad t \geq 0 && (8.14)
\end{aligned}$$

for the unknown displacements u and the unknown order parameter S that will be examined and implemented in the following part.

For the numerical tests in Chapter 10 we will restrict our considerations to the two-dimensional case assuming plane strain, see Appendix 5.

8.1 Variational formulation

In general, the system (8.9)–(8.14) has no classical solution. In order to solve the phase field problem numerically, we need to transform it into a variational formulation, which will be done next.

Let S and u be smooth enough solutions of the system (8.9)–(8.14). For a variational formulation, we multiply Eq. (8.11) with test functions $v_S \in H_0^1$ and integrate over the given domain leading to

$$\int_\Omega h(|\nabla S|)\partial_t S v_S \, dx = -\int_\Omega \big(W_S(S,\varepsilon(u)) + \beta\psi'(S) - \alpha\Delta S\big)\, v_S \, dx \,. \tag{8.15}$$

Applying the product rule, the divergence theorem and homogeneous boundary values for v_S, yields

$$\int_\Omega h(|\nabla S|)\partial_t S v_S \, dx = -\int_\Omega \big(W_S(S,\varepsilon(u)) + \beta\psi'(S)\big)\, v_S \, dx - \alpha \int_\Omega \nabla S \cdot \nabla v_S \, dx \,. \tag{8.16}$$

Next, we test Eq. (8.9) with $v_u \in \left(H_0^1\right)^3$ to have

$$-\int_\Omega (\nabla \cdot T(S,\varepsilon(u))) \cdot v_u \, dx \;=\; \int_\Omega b \cdot v_u \, dx \,. \tag{8.17}$$

With the product rule and the divergence theorem we obtain

$$\begin{aligned}\int_\Omega (\nabla \cdot T(S,\varepsilon(u))) \cdot v_u \, dx &= \int_\Omega \nabla \cdot (T(S,\varepsilon(u))v_u)\, dx - \int_\Omega T(S,\varepsilon(u)) : \nabla v_u \, dx \\ &= \int_{\partial\Omega} (T(S,\varepsilon(u))v_u) \cdot n \, da - \int_\Omega T(S,\varepsilon(u)) : \nabla v_u \, dx \,.\end{aligned}$$

Exploiting the homogeneous Dirichlet boundary condition of v_u, the first term on the right-hand side disappears and Eq. (8.17) transforms to

$$\int_\Omega T(S,\varepsilon(u)) : \nabla v_u \, dx = \int_\Omega b \cdot v_u \, dx \,. \tag{8.18}$$

Using the symmetry of the stress tensor, we calculate

$$\begin{aligned}T(S,\varepsilon(u)) : \nabla v_u &= \frac{1}{2} T(S,u) : \nabla v_u + \frac{1}{2} T(S,u)^\mathsf{T} : \nabla v_u \\ &= \frac{1}{2} T(S,u) : \nabla v_u + \frac{1}{2} T(S,u) : (\nabla v_u)^\mathsf{T}\end{aligned}$$

$$
\begin{aligned}
&= T(S,u) : \frac{1}{2}\left(\nabla v_u + (\nabla v_u)^{\mathsf{T}}\right) \\
&= T(S,u) : \varepsilon(v_u)
\end{aligned}
$$

and receive

$$
\int_\Omega T(S,\varepsilon(u)) : \varepsilon(v_u)\, dx = \int_\Omega b \cdot v_u\, dx\,. \tag{8.19}
$$

Thus, the variational formulations of Eq. (8.9) and Eq. (8.11) are given by

$$
\begin{aligned}
\langle T(S,\varepsilon(u)), \varepsilon(v_u)\rangle - \langle b, v_u\rangle &= 0\,, && (8.20)\\
\langle h(|\nabla S|)\partial_t S, v_S\rangle + \langle W_S(S,\varepsilon(u)) + \beta\psi'(S), v_S\rangle + \alpha\langle \nabla S, \nabla v_S\rangle &= 0, && (8.21)\\
\forall v_S \in H_0^1(\Omega)\,, \forall v_u \in \left(H_0^1(\Omega)\right)^3\,,\ t > 0.
\end{aligned}
$$

The well-posedness and the existence of weak solutions of a system related to (8.20)–(8.21) was studied in [46], [45], [21], [19].

The well-posedness of the elastic hybrid model was studied in [1]–[4], [92], [91] and [87], giving existence proofs for the one-dimensional case. To show the existence and uniqueness of solutions for the hybrid model in two- and three dimensions is an open problem.

8.2 Energy decay

In general, phase transitions can depend on different types of inner energies, such as for example, thermal energy as a disordered movement of molecules, which can be measured by temperature changes. A general approach to show the thermodynamic consistency of a physically motivated model is the verification of the second law of thermodynamics. Since the parameters μ, λ and ν are assumed as temperature-independent in the present work, we refer the interested reader to Appendix 6 and restrict ourselves to the discussion of the energy decay property along the lines of Chapter 5.

In addition to the phase field equation (8.11) we have the equilibrium condition equation (8.9) that has to be considered within the formulation of the respective free energy. Adding the respective term $-b \cdot u$ to the free energy function Eq. (8.1) we have

$$\begin{aligned}\bar{E}_{el,b}(S(t),\varepsilon(u(t))) &= \int_\Omega \frac{1}{2}\left(\varepsilon(u(t)) - \bar{\varepsilon}S(t)\right) : \mathbb{C}(S(t))\left(\varepsilon(u(t)) - \bar{\varepsilon}S(t)\right) - b\cdot u \\ &+ \beta\psi(S(t)) + \frac{\alpha}{2}\left|\nabla S(t)\right|^2\,dx\,. \end{aligned} \tag{8.22}$$

For ease of notation we use S_t for $\partial_t S$ from now an and state

Lemma 9 (*Energy decay for the general model with elasticity*). *Let S and u be sufficiently smooth solutions of the Eq.-system (8.9)-(8.14). Further assume $u_t \in \left(H_0^1(\Omega)\right)^3$. Then the energy function (8.22) satisfies the condition*

$$\frac{d}{dt}\bar{E}_{el,b}(S(t),\varepsilon(u(t))) = -\int_\Omega c\left(|\nabla S(t)|\right)S_t^2\,dx \le 0\,. \tag{8.23}$$

Proof. We calculate

$$\begin{aligned}\frac{d}{dt}\bar{E}_{el,b}(S(t),\varepsilon(u(t))) &= \int_\Omega \frac{d}{dt}\Big(\frac{1}{2}\left(\varepsilon(u(t)) - \bar{\varepsilon}S(t)\right) : \mathbb{C}(S(t))\left(\varepsilon(u(t)) - \bar{\varepsilon}S(t)\right) \\ &- b\cdot u(t) + \beta\,\psi(S(t)) + \frac{\alpha}{2}|\nabla S(t)|^2\Big)\,dx \\ &= \int_\Omega (\varepsilon(u_t) - \bar{\varepsilon}S_t) : \left(\mathbb{C}(S(t))\left(\varepsilon(u(t)) - \bar{\varepsilon}S(t)\right)\right) \\ &+\frac{1}{2}\left(\varepsilon(u(t)) - \bar{\varepsilon}S(t)\right) : \left((\mathbb{C}_2 - \mathbb{C}_1)S_t\left(\varepsilon(u(t)) - \bar{\varepsilon}S(t)\right)\right) \\ &-b\cdot u_t + \beta\,\psi'(S(t))S_t + \alpha\,\nabla S(t)\cdot\nabla S_t\,dx \\ &= \int_\Omega [(-\bar{\varepsilon}) : \mathbb{C}(S(t))\left(\varepsilon(u(t)) - \bar{\varepsilon}S(t)\right) \\ &+\frac{1}{2}\left(\varepsilon(u(t)) - \bar{\varepsilon}S(t)\right) : \left((\mathbb{C}_2 - \mathbb{C}_1)\left(\varepsilon(u(t)) - \bar{\varepsilon}S(t)\right)\right) \\ &+ \beta\,\psi'(S(t))\,]\,S_t + \alpha\,\nabla S(t)\cdot\nabla S_t - b\cdot u_t \\ &+ \varepsilon(u_t) : \left(\mathbb{C}(S(t))\left(\varepsilon(u(t)) - \bar{\varepsilon}S(t)\right)\right)\,dx\,.\end{aligned}$$

Inserting Eq. (8.20) with $v_u = u_t$ yields

$$\begin{aligned}\frac{d}{dt}\bar{E}_{el,b}(S(t),\varepsilon(u(t))) &= \int_\Omega [(-\bar{\varepsilon}) : \left(\mathbb{C}(S(t))\left(\varepsilon(u(t)) - \bar{\varepsilon}S(t)\right)\right) \\ &+\frac{1}{2}\left(\varepsilon(u(t)) - \bar{\varepsilon}S(t)\right) : \left((\mathbb{C}_2 - \mathbb{C}_1)\left(\varepsilon(u(t)) - \bar{\varepsilon}S(t)\right)\right)\end{aligned}$$

$$+ \beta \psi'(S(t)) \,] \, S_t + \alpha \nabla S(t) \cdot \nabla S_t \, dx$$

and using Eq. (8.21) with $v_S = S_t$ leads to

$$\frac{d}{dt} \bar{E}_{el,b}(S(t), \varepsilon(u(t))) = - \int_\Omega h(|\nabla S(t)|) S_t^2 \, dx \leq 0 \, .$$

□

Now, that we have explained the additional terms for the continuous elastic phase field model and shown its energy decay property, we will prepare the numerical implementation by discretising the model in the next chapter.

9 Discretisation of the general model coupled to linear elasticity

Discretisation in space. Along the lines of Chapter 7.1, where we defined the finite dimensional spaces $V_h \subset H^1(\Omega)$ and $V_{h,0} \subset H_0^1(\Omega)$ for the Finite Element method, we define the finite dimensional space $W_h \subset \left(H^1\right)^3$ and its subspaces

$$\begin{aligned} W_{h,0} &= \{ v_{hu} \in W_h | v_{hu} = 0 \quad \text{in } \partial\Omega \} \, , \\ W_{h,D} &= \{ v_{hu} \in W_h | v_{hu} = u_D \text{ in } \partial\Omega \} \, . \end{aligned}$$

Problem (*Variational semi-discrete formulation with elasticity*). Let $S_{h,0} \in V_{h,0}$ be given. Find $S_h \in C^1([0,T]; V_{h,0})$ and $u_h \in C^1([0,T]; W_{h,D})$ such that

$$\langle T(S_h, \varepsilon(u_h)), \varepsilon(v_{hu}) \rangle = \langle b, v_{hu} \rangle \, , \tag{9.1}$$

$$\begin{aligned} \langle h\left(|\nabla S_h|\right) S_{h,t}, v_{hS} \rangle + \langle W_S(S_h, \varepsilon(u_h)), v_{hS} \rangle & \\ + \langle \beta \psi'(S_h), v_{hS} \rangle + \langle \alpha \nabla S_h, \nabla v_{hS} \rangle &= 0 \end{aligned} \tag{9.2}$$

for all discrete test functions $v_{hS} \in V_{h,0}, v_{hu} \in W_{h,0}$ and $t > 0$.

Since the discretised elastic strain ε_h depends on the discretised values S_h and u_h, we replace the arguments of the respective functions by (S_h, u_h) instead of (S_h, ε_h). For the later two-dimensional numerical implementation we define a time independent outer force $b = (b_x, b_y)^\mathsf{T} \in \mathbb{R}^2$.

With definition (8.6), the discrete elastic energy is given by

$$W(S_h, u_h) = \frac{1}{2}(\varepsilon(u_h) - \bar{\varepsilon} S_h)^{\mathsf{T}} : ((\mathbb{C}_1 - S_h(\mathbb{C}_2 - \mathbb{C}_1))\,(\varepsilon(u_h) - \bar{\varepsilon} S_h))\,, \tag{9.3}$$

$$\varepsilon(u_h) = \frac{1}{2}\left(\nabla u_h + (\nabla u_h)^{\mathsf{T}}\right) \tag{9.4}$$

and the discrete stress tensor yields

$$T(S_h, u_h) = \mathbb{C}(S_h)(\varepsilon(u_h) - \bar{\varepsilon} S_h)\,. \tag{9.5}$$

First, we prove the

Lemma 10 (*Energy decay for the time discretisation for the elastic model*)**.** *Let $S_h \in C^1([0,T]; V_{h,0})$ be a solution of Eq. (9.2) and let $u_h \in C^1([0,T]; W_h) \cap C^0([0,T]; W_{h,D})$ be a solution of Eq. (9.1). Further assume $u_{h,t} \in C^0([0,T]; W_{h,0})$. Then the free energy function satisfies the condition*

$$\frac{d}{dt}\bar{E}_{el,b}(S_h, u_h) = -\int_\Omega h\,(|\nabla S_h|)\,S_{h,t}^2\,dx \le 0\,. \tag{9.6}$$

Proof. The energy decay is verified by discretising and deriving Eq. (8.22). We insert Eq. (9.2) with $v_{hS} = S_{h,t}$ and Eq. (9.1) with $v_{hu} = u_{h,t}$ and this leads to

$$\begin{aligned}
\frac{d}{dt}\bar{E}_{el,b}(S_h, u_h) &= \int_\Omega [(-\bar{\varepsilon}) : \mathbb{C}(S_h)\,(\varepsilon(u_h) - \bar{\varepsilon} S_h) \\
&\quad + \frac{1}{2}\,(\varepsilon(u_h) - \bar{\varepsilon} S_h) : ((\mathbb{C}_2 - \mathbb{C}_1)\,(\varepsilon(u_h) - \bar{\varepsilon} S_h)) \\
&\quad + \beta\,\psi'(S_h)\,]\,S_{h,t} - \alpha\,\nabla S_h \cdot \nabla S_{h,t} - b \cdot u_{h,t} \\
&\quad + \varepsilon(u_{h,t}) : (\mathbb{C}(S_h)\,(\varepsilon(u_h) - \bar{\varepsilon} S_h))\,dx \\
&= -\int_\Omega h(|\nabla S_h|) S_{h,t}^2 \;\le\; 0\,.
\end{aligned}$$

□

Now we look at the well-posedness of the semi-discretised phase field equation. For our theoretical considerations and the following numerical implementations we uncouple Eq. (9.1) and Eq. (9.2). In the numerical simulations we will first solve Eq. (9.1) and insert the calculated value for the displacements u_h at the timestep n into Eq. (9.2). We will use this approach also in our theoretical numerical considerations in the following and formulate the

Assumption 4. *$W(S)$ is twice continuously differentiable with respect to S, $W(S) \geq C_{W,0}|S| - C_{W,1}$ and $|W_{SS}(S)| \leq C_{W,2}$ $\forall S$ with $C_{W,i}, \in \mathbb{R}_0^+,\ i = 0, 2$.*

The existence of u_h will be discussed in the next section within the fully discretised formulation and we formulate

Lemma 11 (*Existence and uniqueness of the semi-discrete formulation with elasticity*). *Let $S_{h,0} \in V_{h,0}$ be given and let $u_h \in W_{h,D}$ be known. Then we have an existing unique solution $S_h \in C^1([0,1]; V_{h,0})$ for Eq. (9.2).*

Proof. The variational form Eq. (9.2) is the same as Eq. (6.1) with an additional term $\langle W_S(S_h, \varepsilon(u)), v_S \rangle$. With Assumption 4, the Lipschitz continuity of the right hand side of Eq. (6.7) is given and therefore we argue exactly as in Chapter 6 to justify the local existence of the solution of S_h.

The boundedness of $||S_h||_{H^1(\Omega)}$ is also shown in a similar way along the lines of Chapter 6.

Therefore, the existence of a global solution is given. □

Discretisation in time. In the next chapter, we validate the Allen-Cahn model and the hybrid model numerically, related to linear elasticity. For this purpose, an appropriate fully-discrete scheme is needed. The formulations Eq. (9.1) and Eq. (9.2) will now be discretised in time and we address some issues on the well-posedness. Further, the energy decay properties of the fully discretised phase field model will be shown.

We define the approximation of the time derivative of the discrete order parameter by

$$S_{h,t} = \frac{S_h^n - S_h^{n-1}}{\tau}. \tag{9.7}$$

The difference quotients for the double well potential and the elastic energy with respect to the order parameter are given by Eq. (6.13) and

$$F(S_h^n, S_h^{n-1}, u_h^n) := \begin{cases} \dfrac{W(S_h^n, u_h^n) - W(S_h^{n-1}, u_h^n)}{S_h^n - S_h^{n-1}} & \text{for } S_h^n \neq S_h^{n-1} \\ W_S(S_h^n) & \text{for } S_h^n = S_h^{n-1}. \end{cases} \tag{9.8}$$

As in Chapter 6, these definition are needed for the proofs of the energy decay property and on the well-posedness of the elastic phase field problem. The formulations chosen within these proofs are implemented in the same way to guarantee confidence in the results.

Thus, with $h(|\nabla S_h^{n-1}|)$ known from the last time step the fully-discrete phase field model can be formulated as

Problem (*Variational fully-discrete formulation with elasticity*). Let $S_{h,0} \in V_{h,0}$ be given. For any $n \geq 1$ find $S_h^n \in V_{h,0}$ and $u_h^n \in W_{h,D}$ such that

$$\langle T(S_h^{n-1}, u_h^n), \varepsilon(v_{hu})\rangle = \langle b, v_{hu}\rangle \,, \tag{9.9}$$

$$\begin{aligned} \langle h(|\nabla S_h^{n-1}|)\frac{S_h^n - S_h^{n-1}}{\tau}, v_{hS}\rangle &+ \langle\frac{W(S_h^n, u_h^n) - W(S_h^{n-1}, u_h^n)}{S_h^n - S_h^{n-1}}, v_{hS}\rangle \\ + \langle\beta\frac{\psi(S_h^n) - \psi(S_h^{n-1})}{S_h^n - S_h^{n-1}}, v_{hS}\rangle &+ \langle\alpha\nabla S_h^n, \nabla v_{hS}\rangle = 0 \end{aligned} \tag{9.10}$$

for all discrete test functions $v_{hS} \in V_{h,0}$ and $v_{hu} \in W_{h,0}$.

To confirm the following numerical implementation, we want the fully-discrete energy to decay. Therefore, we need to define the stress tensor and the test function in exactly the following way

$$T(S_h^{n-1}, u_h^n) = \mathbb{C}(S_h^{n-1})\tilde{\varepsilon}(S_h^{n-1}, u_h^n)\,, \tag{9.11}$$

$$v_{hu} = 2(u_h^n - u_h^{n-1})\,. \tag{9.12}$$

Now we can examine the energy decay property of the system (9.9)-(9.10). Defining $\bar{E}^n_{el,u_h} = \bar{E}_{el,b}(S_h^n, u_h^n)$ and using the definition (6.16) we state

Lemma 12 (*Energy decay for the fully discretised general model with elasticity*). *Let $S_h^n \in V_{h,0}$ $\forall t$ be a solution of Eq. (9.10) and let $u_h^n \in W_{h,D}$ $\forall t$ be a solution of Eq. (9.9). Then the discrete free energy satisfies the condition*

$$\bar{\partial}_\tau \bar{E}_{el,uh} \leq 0\,. \tag{9.13}$$

Proof. With the discretised free energy (8.22)

$$\bar{E}^n_{el,uh} = \int_\Omega W(S_h^n, u_h^n) + \beta\psi(S_h^n) + \frac{\alpha}{2}|\nabla S_h^n|^2 - b \cdot u_h^n \, dx$$

and

$$W(S_h^n, u_h^n) = \frac{1}{2}\left(\varepsilon(u_h^n) - \bar{\varepsilon} S_h^n\right) : \mathbb{C}(S_h^n)\left(\varepsilon(u_h^n) - \bar{\varepsilon} S_h^n\right) \tag{9.14}$$

we calculate

$$\begin{aligned}
\frac{\bar{E}_{el,uh}^n - \bar{E}_{el,uh}^{n-1}}{\tau} &= \frac{1}{\tau}\int_\Omega W(S_h^n, u_h^n) - W(S_h^{n-1}, u_h^{n-1}) \\
&- b(u_h^n - u_h^{n-1}) + \beta\left(\psi(S_h^n) - \psi(S_h^{n-1})\right) \\
&+ \alpha\nabla S_h^n \cdot (\nabla S_h^n - \nabla S_h^{n-1}) - \frac{\alpha}{2}\left|\nabla S_h^n - \nabla S_h^{n-1}\right|^2 dx\,.
\end{aligned}$$

For the last two terms, see Appendix 7.

This equation can be transformed to

$$\begin{aligned}
\frac{\bar{E}_{el,uh}^n - \bar{E}_{el,uh}^{n-1}}{\tau} &= \frac{1}{\tau}\int_\Omega W(S_h^n, u_h^n) - W(S_h^{n-1}, u_h^n) + W(S_h^{n-1}, u_h^n) \\
&- W(S_h^{n-1}, u_h^{n-1}) - b(u_h^n - u_h^{n-1}) + \beta\left(\psi(S_h^n) - \psi(S_h^{n-1})\right) \\
&+ \alpha\nabla S_h^n \cdot (\nabla S_h^n - \nabla S_h^{n-1}) - \frac{\alpha}{2\tau}\left|\nabla S_h^n - \nabla S_h^{n-1}\right|^2 dx \\
&= \frac{1}{\tau}\int_\Omega \frac{W(S_h^n, u_h^n) - W(S_h^{n-1}, u_h^n)}{S_h^n - S_h^{n-1}}(S_h^n - S_h^{n-1}) \\
&+ \beta\frac{\psi(S_h^n) - \psi(S_h^{n-1})}{S_h^n - S_h^{n-1}}(S_h^n - S_h^{n-1}) \\
&+ \alpha\nabla S_h^n \cdot (\nabla S_h^n - \nabla S_h^{n-1}) - \frac{\alpha}{2\tau}\left|\nabla S_h^n - \nabla S_h^{n-1}\right|^2 \\
&+ \frac{1}{\tau}\int_\Omega \underbrace{W(S_h^{n-1}, u_h^n) - W(S_h^{n-1}, u_h^{n-1}) - b(u_h^n - u_h^{n-1})}_{**}\, dx\,.
\end{aligned}$$

Using Eq. (9.14) and Eq. (8.4) leads to

$$\begin{aligned}
** &= \frac{1}{2}\tilde{\varepsilon}(S_h^{n-1}, u_h^n) : \left(\mathbb{C}(S_h^{n-1})\tilde{\varepsilon}(S_h^{n-1}, u_h^n)\right) \\
&- \frac{1}{2}\tilde{\varepsilon}(S_h^{n-1}, u_h^{n-1}) : \left(\mathbb{C}(S_h^{n-1})\tilde{\varepsilon}(S_h^{n-1}, u_h^{n-1})\right) \\
&- \frac{1}{2}\tilde{\varepsilon}(S_h^{n-1}, u_h^n) : \left(\mathbb{C}(S_h^{n-1})\tilde{\varepsilon}(S_h^{n-1}, u_h^{n-1})\right) \\
&+ \frac{1}{2}\tilde{\varepsilon}(S_h^{n-1}, u_h^{n-1}) : \left(\mathbb{C}(S_h^{n-1})\tilde{\varepsilon}(S_h^{n-1}, u_h^n)\right) \\
&= \frac{1}{2}\tilde{\varepsilon}(S_h^{n-1}, u_h^n) : \left(\mathbb{C}(S_h^{n-1})(\tilde{\varepsilon}(S_h^{n-1}, u_h^n) - \tilde{\varepsilon}(S_h^{n-1}, u_h^{n-1})\right) + \\
&+ \frac{1}{2}\tilde{\varepsilon}(S_h^{n-1}, u_h^{n-1}) : \left(\mathbb{C}(S_h^{n-1})\left(\tilde{\varepsilon}(S_h^{n-1}, u_h^n) - \tilde{\varepsilon}(S_h^{n-1}, u_h^{n-1})\right)\right)
\end{aligned}$$

$$
\begin{aligned}
&= \frac{1}{2}\left(\tilde{\varepsilon}(S_h^{n-1}, u_h^n) + \tilde{\varepsilon}(S_h^{n-1}, u_h^{n-1})\right) : \left(\mathbb{C}(S_h^{n-1})\left(\tilde{\varepsilon}(S_h^{n-1}, u_h^n) - \tilde{\varepsilon}(S_h^{n-1}, u_h^{n-1})\right)\right) \\
&= \tilde{\varepsilon}(S_h^{n-1}, u_h^n) : \left(\mathbb{C}(S_h^{n-1})(\varepsilon(u_h^n - u_h^{n-1}))\right) + \\
&- \frac{1}{2}\left(\varepsilon(u_h^n - u_h^{n-1}) : \left(\mathbb{C}(S_h^{n-1})(\varepsilon(u_h^n - u_h^{n-1}))\right)\right) .
\end{aligned}
$$

For the last term, see Appendix 7. So we have

$$
\begin{aligned}
\frac{\bar{E}_{el,uh}^n - \bar{E}_{el,uh}^{n-1}}{\tau} &= \frac{1}{\tau}\int_\Omega \tilde{\varepsilon}(S_h^{n-1}, u_h^n) : \left(\mathbb{C}(S_h^{n-1})(\varepsilon(u_h^n - u_h^{n-1}))\right) \\
&- b \cdot (u_h^n - u_h^{n-1}) + \frac{W(S_h^n, u_h^n) - W(S_h^{n-1}, u_h^n)}{S_h^n - S_h^{n-1}}(S_h^n - S_h^{n-1}) \\
&+ \beta\frac{\psi(S_h^n) - \psi(S_h^{n-1})}{S_h^n - S_h^{n-1}}(S_h^n - S_h^{n-1}) + \alpha \nabla S_h^n \cdot (\nabla S_h^n - \nabla S_h^{n-1}) \\
&- \frac{1}{2}\left(\varepsilon(u_h^n - u_h^{n-1}) : \left(\mathbb{C}(S_h^{n-1})(\varepsilon(u_h^n - u_h^{n-1}))\right)\right) dx \\
&- \frac{\alpha}{2\tau}||\nabla S_h^n - \nabla S_h^{n-1}||_{L_2(\Omega)}^2 .
\end{aligned} \tag{9.15}
$$

Applying Eq. (9.9) with $v_{hu} = u_h^n - u_h^{n-1}$ we can drop the first two terms on the right-hand side and have

$$
\begin{aligned}
\frac{\bar{E}_{el,uh}^n - \bar{E}_{el,uh}^{n-1}}{\tau} &= \frac{1}{\tau}\int_\Omega \frac{W(S_h^n, u_h^n) - W(S_h^{n-1}, u_h^n)}{S_h^n - S_h^{n-1}}(S_h^n - S_h^{n-1}) \\
&+ \beta\frac{\psi(S_h^n) - \psi(S_h^{n-1})}{S_h^n - S_h^{n-1}}(S_h^n - S_h^{n-1}) + \alpha \nabla S_h^n \cdot (\nabla S_h^n - \nabla S_h^{n-1})\, dx \\
&- \frac{1}{2}\int_\Omega \left(\varepsilon(u_h^n - u_h^{n-1}) : \left(\mathbb{C}(S_h^{n-1})(\varepsilon(u_h^n - u_h^{n-1}))\right)\right) dx \\
&- \frac{\alpha}{2\tau}||\nabla S_h^n - \nabla S_h^{n-1}||_{L_2(\Omega)}^2
\end{aligned}
$$

and inserting $v_{hS} = (S_h^n - S_h^{n-1})$ in Eq. (9.10) yields

$$
\begin{aligned}
\frac{\bar{E}_{el,uh}^n - \bar{E}_{el,uh}^{n-1}}{\tau} &= -\frac{1}{\tau}\int_\Omega \frac{1}{h(|\nabla S_h^n|)}(S_h^n - S_h^{n-1})^2\, dx \\
- \frac{1}{2\tau}\int_\Omega \left(\varepsilon(u_h^n - u_h^{n-1}) : \left(\mathbb{C}(S_h^{n-1})(\varepsilon(u_h^n - u_h^{n-1}))\right)\right) dx &- \frac{\alpha}{2\tau}||\nabla S_h^n - \nabla S_h^{n-1}||_{L_2(\Omega)}^2 \leq 0 .
\end{aligned}
$$

□

Before we start to explain the implementation, we will at last examine the well-posedness of the fully discretised uncoupled elastic phase field system and state

Lemma 13 (*Existence and uniqueness of the fully discretised system with elasticity*). *Let $S_h^{n-1} \in V_{h,0}$ be given. Then the uncoupled system (9.9)–(9.10) has a unique solution $S_h^n \in V_{h,0}$ and $u_h^n \in W_{h,D}$ for all time steps $0 < \tau < \tau_0$ with τ_0 small enough.*

Proof. Since we solve Eq. (9.9) and Eq. (9.10) consecutive, we start with the uniqueness and existence of the solution u_h^n. The Finite Element discretisation of Eq. (9.9) leads to the formulation (A.4.9) in Appendix 4.2. This equation can be expressed as

$$B(u_h, v_{hu}) = \langle \bar{b}, v_{hu} \rangle - B(u_D, v_{hu}), \tag{9.16}$$

minimising the functional

$$J(u_h) := \frac{1}{2} B(u_h, v_{hu}) - l(\tilde{b}, v_{hu}) \tag{9.17}$$

with

$$\bar{b} = b + B_J^{u\top} \mathbb{C}(N_I \hat{S}_I) \hat{S}_J \bar{\varepsilon}, \tag{9.18}$$

$B(u_D, v_{hu})$ adjusting $B(u_h, v_{hu})$ to a homogeneous boundary condition in order to have $u_h \in H_0^1$ at the left hand side of Eq. (9.16) and $\tilde{b}$ for a more compact notation of the right-hand side of Eq. (9.16). With $L(v_{hu}) = l(\tilde{b}, v_{hu}), L : H_0^1 \to \mathbb{R}$ we can rewrite Eq. (9.16) as

$$L(v_{hu}) = B(u_h, v_{hu}) \quad \forall v_{hu} \in H_0^1(\Omega). \tag{9.19}$$

This problem was studied in [30] with the result that for the ellipticity and the continuity of $B : H_0^1 \times H_0^1 \to \mathbb{R}$ in Eq. (9.19) the Riesz representation theorem garanties a unique solution for u_h, assuming $\mathbb{C}(S)$ being positive-definite. For more details see also [23] and [28].

With the existing, given u_h^n we can now examine Eq. (9.10). With the abbreviations Eq. (9.8) and Eq. (6.13) the discrete variational formulation reads

$$\begin{aligned} \langle h(|\nabla S_h^{n-1}|) \frac{S_h^n - S_h^{n-1}}{\tau}, v_{hS} \rangle \quad &+ \quad \langle F(S_h^n, S_h^{n-1}, u_h^n), v_{hS} \rangle \\ +\beta \langle D(S_h^n, S_h^{n-1}), v_{hS} \rangle \quad &+ \quad \alpha \langle \nabla S_h^n, \nabla v_{hS} \rangle = 0. \end{aligned} \tag{9.20}$$

Same as Eq. (7.9), Eq. (9.20) can be formulated as fixed-point equation

$$\begin{aligned} \langle S_h^n, v_{hS} \rangle = & - \frac{\tau}{h(|\nabla S_h^{n-1}|)} \langle F(S_h^n, S_h^{n-1}, u_h^n), v_{hS} \rangle - \frac{\beta}{h(|\nabla S_h^{n-1}|)} \tau \langle D(S_h^n, S_h^{n-1}), v_{hS} \rangle \\ & - \frac{\alpha}{h(|\nabla S_h^{n-1}|)} \tau \langle \nabla S_h^n, \nabla v_{hS} \rangle + \langle S_h^{n-1}, v_{hS} \rangle \,. \end{aligned} \tag{9.21}$$

Using Assumption 4 and knowing u_h^n from the time step before, we can adapt our proof of the existence of the solution of S_h^n of Chapter 6. The fully discretised elastic term $F(S_h^n, S_h^{n-1}, u_h^n)$ in Eq. (9.20) is structured like the double well term $D(S_h^n, S_h^{n-1})$. As shown in Eq. (7.15) we can perform a polynomial division for $F(S_h^n, S_h^{n-1}, u_h^n)$ leading to a polynom of second order in S_h^n.

We replace the expression $\psi(S_h^n) - \psi(S_h^{n-1})$ by

$$\psi_W(S_h^n) - \psi_W(S_h^{n-1}) := \psi(S_h^n) - \psi(S_h^{n-1}) + W(S_h^n) - W(S_h^{n-1}) \tag{9.22}$$

and assume that $\psi_W(S_h^n)$ fulfills the conditions of Assumption 3.

Thus, the proof goes along the lines of the proof in chapter 6 and we have the existence of a solution of S_h^n.

The uniqueness of the solution S_h^n can also be shown analogously, expanding the proof in Chapter 6 to the elasticity term, premising that the displacements u_h^n are given.

Adding the term $\langle F(S_h^n, S_h^{n-1}) - F(\tilde{S}_h^n, S_h^{n-1}), S_h^n - \tilde{S}_h^n \rangle$ on the right-hand side of Eq. (6.30), we repeat the proof of Lemma 8. With Assumption 4 and applying Cauchy-Schwarz we get

$$\begin{aligned} \frac{h}{\tau} ||S_h^n - \tilde{S}_h^n||_{L_2(\Omega)}^2 & + \alpha ||\nabla S_h^n - \nabla \tilde{S}_h^n||_{L_2(\Omega)}^2 \\ & \leq ||F(S_h^n, S_h^{n-1}) - F(\tilde{S}_h^n, S_h^{n-1})||_{L_2(\Omega)} ||S_h^n - \tilde{S}_h^n||_{L_2(\Omega)} \\ & + \beta ||D(S_h^n, S_h^{n-1}) - D(\tilde{S}_h^n, S_h^{n-1})||_{L_2(\Omega)} ||S_h^n - \tilde{S}_h^n||_{L_2(\Omega)} \,. \end{aligned} \tag{9.23}$$

If we know the displacements u_h^n, we apply the fundamental theorem of calculus as in Lemma 8 on the respective terms, leading to

$$|F(S_h^n, S_h^{n-1}) - F(\tilde{S}_h^n, S_h^{n-1})| + \beta |D(S_h^n, S_h^{n-1}) - D(\tilde{S}_h^n, S_h^{n-1})| \leq (\beta \bar{d} + \bar{f}) |S_h^n - \tilde{S}_h^n| \,. \tag{9.24}$$

Combining Eq. (9.23) with Eq. (9.24) yields

$$\underline{h}||S_h^n - \tilde{S}_h^n||^2_{L_2(\Omega)} + \alpha\tau||\nabla S_h^n - \nabla\tilde{S}_h^n||^2_{L_2(\Omega)} \leq (\beta\bar{\tilde{d}} + \bar{\tilde{f}})\tau||S_h^n - \tilde{S}_h^n||^2_{L_2(\Omega)} \tag{9.25}$$

and can be transformed to

$$\left(\frac{\underline{h}}{\tau} - (\beta\bar{\tilde{d}} + \bar{\tilde{f}})\right)||S_h^n - \tilde{S}_h^n||^2_{L_2(\Omega)} + \alpha||\nabla S_h^n - \nabla\tilde{S}_h^n||^2_{L_2(\Omega)} \leq 0. \tag{9.26}$$

If we choose τ small enough, the only way to satisfy the inequality is to define $S_h^n = \tilde{S}_h^n$, which proves the uniqueness of S_h^n. □

10 Numerical validation

We have confirmed the well-posedness of the fully discretised formulation of the general uncoupled elastic phase field system (9.9) - (9.10) and we have proven that the free energy decays along discrete solutions. This allows us to implement the models into a numerical environment in order to carry out some exemplary tests.

10.1 Ersatzstress-Term

First, we show a numerical example from [4]. For a better understanding, we summarise some of the details in [4]. Instead of solving the coupled linear elasticity system, the authors derive a term we call "ersatzstress". Again we use the nomenclature of the original literature, and thus, the elasticity tensor is denoted by D.

The following terms are the result of studying the original literature and discussions with the author. The explanations should give an idea of what is simulated in this chapter.

In [4], the stress tensor is splitted into a first part independent of the boundary value problem of elasticity and a second part that depends on it

$$T = [\hat{T}]S + w\,. \tag{10.1}$$

The elastic energy enters the phase field equation via the term

$$W_{S,ers} = -T : \bar{\varepsilon}\,. \tag{10.2}$$

The stress component w is evaluated as a one-dimensional solution of a boundary problem of linear elasticity and given as a constant value, see [4] for more details. The term $[\hat{T}]$ describes the jump of the stresses at the sharp interface and can be expressed by

$$\begin{aligned} [\hat{T}] &= (D(\varepsilon - \bar{\varepsilon}S))^{(-)} - (D(\varepsilon - \bar{\varepsilon}S))^{(+)} \\ &= D(\varepsilon^{(-)} - \varepsilon^{(+)} - \bar{\varepsilon}) \\ &= D\bar{\varepsilon}(P_n - 1)\,. \end{aligned} \tag{10.3}$$

The last step is based on the equation

$$P_n\bar{\varepsilon} = [\varepsilon(\nabla_x \hat{u})]$$

with the orthogonal projection P_n. Inserting Eq. (10.1) and Eq. (10.3) into Eq. (10.2) yields

$$W_{S,ers} = -L_1 S - si \tag{10.4}$$

$$\text{with} \qquad L_1 = \bar{\varepsilon} : D(P_n - 1)\bar{\varepsilon}, \quad si = \bar{\varepsilon} : w\,. \tag{10.5}$$

The calculation of L_1 is based on the assumption that the eigenstrain is a multiple of the identity tensor

$$\bar{\varepsilon} = d_1\,\mathrm{I}, \quad d_1 \in \mathbb{R} \tag{10.6}$$

and that the material is isotropic

$$D\sigma = \nu_1\sigma + \nu_2\,\mathrm{trace}\,(\sigma)\mathrm{I}\,. \tag{10.7}$$

The calculation in [4] leads to

$$L_1 = d_1^2(\nu_1 + 3\nu_2)\left(\frac{\nu_1 + 3\nu_2}{\nu_1 + \nu_2} - 3\right) \leq 0, \quad \nu_1 > 0, \;\; \nu_1 + 3\nu_2 > 0\,. \tag{10.8}$$

Insertion of Eq. (10.4) into Eq. (8.2) gives the general ersatzstress model in terms of

$$h(|\nabla S|)\partial_t S = L_1 S + si - \beta\psi'(S) + \alpha\Delta S\,. \tag{10.9}$$

To compare the solution of this model applied to a shrinking circle as in Chapter 7.2, we need to adjust the normal interface velocities Eq. (5.14) and Eq. (5.15). For the jump of the Eshelby tensor we have

$$n \cdot [\hat{C}]n = [\hat{E}(S, \varepsilon)] - \langle\hat{T}\rangle : \bar{\varepsilon}\,, \tag{10.10}$$

see [71]. The energy term $\hat{E}$ is continuous over the sharp interface so that its jump is zero. The mean value

$$\langle \hat{T} \rangle := \frac{1}{2}\left(T^{(-)} + T^{(+)}\right) \tag{10.11}$$

and the jump of the stresses at the sharp interface

$$[\hat{T}] := T^{(+)} - T^{(-)} \tag{10.12}$$

$$\begin{aligned} \text{lead to} \quad n[\hat{C}]n &= -\langle \hat{T} \rangle : \bar{\varepsilon} \\ &= -\frac{1}{2}[\hat{T}] : \bar{\varepsilon} - T^{(-)} : \bar{\varepsilon} \\ &= -\frac{1}{2}\bar{\varepsilon} D(I - P_n) : \bar{\varepsilon} - T^{(-)} : \bar{\varepsilon} \\ &= -\frac{1}{2}L_1 - si\,. \end{aligned} \tag{10.13}$$

Thus, the normal interface velocities (5.14) and (5.15), choosing $c = 1$ and $\tilde{c} = c_1$, with the ersatzstress term read

$$s_{AC,ers} = -\frac{1}{2}L_1 - si + \lambda^{1/2} c_1 \kappa_\Gamma + O(\mu^{1/2}) \tag{10.14}$$

for the Allen-Cahn model and

$$s_{H,ers} = -\frac{1}{2}L_1 - si + \nu^{1/2} \omega_1 \kappa_\Gamma \tag{10.15}$$

for the hybrid model.

We start a simulation to validate this example with $\nu_1 = \nu_2 = 20$ and $d_1 = 0.05$, yielding a value of $L_1 = -0.2$. The value si depends on the boundary condition and is set to $si = -0.3$. So we have $n \cdot [\hat{C}]n = 0.4$ and define the solution of the shrinking circle by solving

$$r_{t_{AC,ers}} = 0.4 + \lambda^{1/2} c_1 r^{-1} \quad \text{and} \quad r_{t_{H,ers}} = 0.4 + \nu^{1/2} \omega_1 r^{-1}\,, \tag{10.16}$$

omitting the error term $O(\mu^{1/2})$ in Eq. (10.14).

There is no classical solution in general for Eq. (10.16) as we had for Eq. (7.22) and therefore, we use a MATLAB ode23 solution algorithm causing a negligible additional numerical error. The obtained numerical solution is compared to the shrinking circle simulated by the phase formulation Eq. (10.9) in terms of

$$h(|\nabla S|)\partial_t S = -0.2S - 0.3 - \beta\psi'(S) + \alpha\Delta S\,. \tag{10.17}$$

The implementation method follows the steps in Chapter 7.1 also explained in Appendix 4. We insert the above calculated values of L_1 and si into Eq. (10.4) and add the resulting expression

$$W_{S,ers} = 0.2S + 0.3 \tag{10.18}$$

to the derivative of the double well potential multiplied with the respective prefactor.

To preserve the double well character of the energy function depending on the sum of the first, second and third term at the right-hand side of Eq. (10.17), we need to confirm the double well character of the functions

$$e_{AC} := 4S^2(1-S)^2 - 0.1S^2 - 0.3S \tag{10.19}$$

for the hybrid model and

$$e_H := \frac{4}{\mu^{1/2}}S^2(1-S)^2 - 0.1S^2 - 0.3S \tag{10.20}$$

for the Allen-Cahn model, which is confirmed, if we choose $\nu = \mu \leq 1$.

Fig. 10.1 shows the simulation results (in blue) of Eq. (10.17) based on the domain Ω and the initial conditions (7.20) and (7.21). For a sufficient small time step of $dt = 0.001$ the hybrid model and the Allen-Cahn model are compared to the numerical solutions Eq. (10.16) (in red). The mesh relation is based upon Eq. (5.29) with $re_A \approx 3$. For this reason, we refine the mesh of the Allen-Cahn model by a factor 3. In contrast to the numerical simulations in Chapter 7, the circle grows due to a different boundary term within Eq. (10.18).

The agreement shown in Fig. 10.1 is sufficient for us at this point, since we only wanted a correspondence in principle with the given examples in [4]. We omit more detailed investigations for this simulation and show further examples of a coupled elasticity system in Chapter 10.3 and Chapter 10.4.

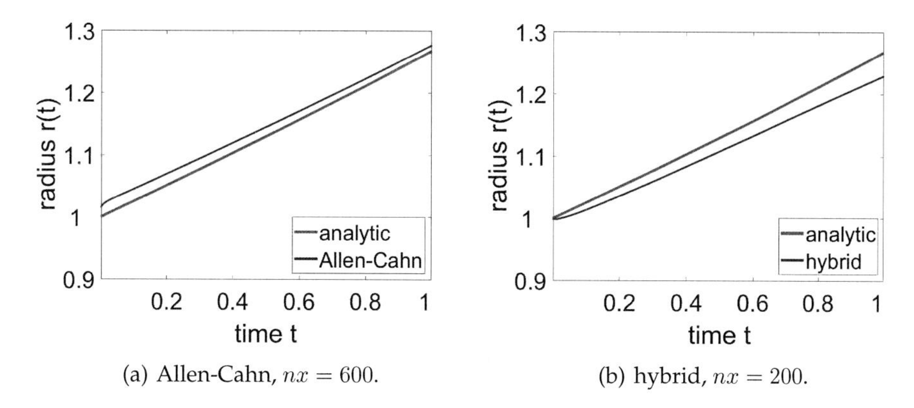

(a) Allen-Cahn, $nx = 600$.

(b) hybrid, $nx = 200$.

Figure 10.1: Comparison of the Allen-Cahn and the hybrid model with the elastic ersatz stress from [4], choosing $\nu = \mu = \lambda = 0.1$.

10.2 Implementation

The implementation of the variational form Eq. (9.9) is based on the definition of the stress tensor in Eq. (9.11). The unknown u_h^n in Eq. (9.9) can be calculated in one step by solving a linear system with known values of S_h^{n-1}, calculated in the former time step.

The implementation of the fully discretised variational formulation (9.10) follows the steps explained in Appendix 4.1 and 4.2, adapting the coefficients of the polynoms $p\left(a_i(\hat{S}_I)^i\right)$ in Eq. (A.4.4) to the extension of the energy expressions $\psi' + W_S$ instead of ψ'. We will show below that $W_S(S_h^n, u_h^n)$ is structured like the double well potential and therefore can be included in the force F on the right-hand side as in Eq. (7.13) and Eq. (7.14).

The implementation scheme processes in the following way.

Step 1: solve

$$\hat{u} = \left(\bar{K}\right)^{-1} \hat{f} \qquad \in \mathbb{R}^{2N} \tag{10.21}$$

for the time step n. The relation arises from the Finite Element discretisation of Eq. (9.10) and refers to the expression

$$u_h(x) = \sum_{I=1}^{N} N_I^u(x)\hat{u}_I \qquad \in \mathbb{R}^2 \tag{10.22}$$

with the nodal unknowns $\hat{u}_I$, which are summarised in $\hat{u}$. Eq. (10.21) contains the elastic global stiffness matrix $\bar{K}$ and the vector of the given forces $\hat{f}$. The right-hand side depends on the order parameter S_h^n of the current time step. For the first time step the initial condition $S_{h,0}$ is inserted. The derivation of the terms involved is explained in Appendix 4.2. The relation is linear and is solved in one step yielding u_h^n.

Step 2: With the determined u_h^n we solve S_h^n for the variational discrete formulation Eq. (9.10). The application of the Banach fixed-point theorem leads to a fixed-point equation

$$S_h^n = \Phi(S_h^n)$$

to be iterated in the current time step n.

The scheme of implementation is shown in Table 10.6. For details see the derivation of Eq. (7.19) and Appendix 4 with an additional entry of the elastic energy W_S in the force F, defined as F_{el}, on the right-hand side. We will show below that this additional entry does not change the structure of F_{el} compared to F. The unknown displacements in W_S are taken from step 1.

Table 10.6: Implicit scheme for elasticity.

Initialisation: $\hat{S}^1 = S(0,x)$, tol=1e-2, err=1		
Loop n=1,Nt		
	Calculate $\bar{K}$ and $\hat{f}$ Solve $\hat{u}^n = \bar{K}^{-1}\hat{f}$	
	$k=-1$, $\hat{S}_{k+1} = \hat{S}^n$	
		While err $>$ tol
		$k = k+1$
		Solve $\hat{S}_{k+1} = K^{-1}_{mod}(\hat{S}^n)\left(M_{2_{mod}}(\hat{S}^n)\hat{S}^n - \beta_T M_1\, F_{el}(\hat{S}_k, \hat{S}^n, \hat{u}^n)\right)$
		err $= \dfrac{\lVert\hat{S}_{k+1} - \hat{S}_k\rVert\max}{\lVert\hat{S}_{k+1}\rVert\max}$
		end
	$\hat{S}^n = \hat{S}_{k+1}$	
end		

Implementation of the elastic energy term The free energy term was expanded to the elastic energy W and instead of $F = \psi'$, as in Chapter 7.1, we calculate

$$F_{el}(S_h^n) = W_S(S_h^n, u_h^n) + \alpha\psi'(S_h^n)\,. \tag{10.23}$$

We use Eq. (7.15) for the derivative of the discrete double well potential.

The additional term $W_S(S_h^n, u_h^n)$ can be calculated in the following way. In the variational formulation Eq. (9.10) the derivative of the elastic energy term is defined by

$$W_S(S_h^n, u_h^n) = \frac{W(S_h^n, u_h^n) - W(S_h^{n-1}, u_h^n)}{S_h^n - S_h^{n-1}}\,. \tag{10.24}$$

We calculate the denominator of the elastic energy term as

$$\begin{aligned}
&W(S_h^n, u_h^n) - W(S_h^{n-1}, u_h^n) = * \\
* &= \frac{1}{2}\tilde{\varepsilon}(S_h^n, u_h^n) : (\mathbb{C}(S_h^n)\tilde{\varepsilon}(S_h^n, u_h^n)) \\
&- \frac{1}{2}\tilde{\varepsilon}(S_h^{n-1}, u_h^n) : \left(\mathbb{C}(S_h^{n-1})\tilde{\varepsilon}(S_h^{n-1}, u_h^n)\right) \\
&= \frac{1}{2}\,[\varepsilon(u_h^n) : (\mathbb{C}(S_h^n)\varepsilon(u_h^n)) - \bar{\varepsilon}S_h^n : (\mathbb{C}(S_h^n)\varepsilon(u_h^n)) \\
&+ \varepsilon(u_h^n) : (\mathbb{C}(S_h^n)\bar{\varepsilon}(-S_h^n)) + \bar{\varepsilon}(-S_h^n) : (\mathbb{C}(S_h^n)\bar{\varepsilon}(-S_h^n))] \\
&- \frac{1}{2}\,\left[\varepsilon(u_h^n) : \left(\mathbb{C}(S_h^{n-1})\varepsilon(u_h^n)\right) - \bar{\varepsilon}S_h^{n-1} : \left(\mathbb{C}(S_h^{n-1})\varepsilon(u_h^n)\right)\right. \\
&+ \left.\varepsilon(u_h^n) : \left(\mathbb{C}(S_h^{n-1})\bar{\varepsilon}(-S_h^{n-1})\right) + \bar{\varepsilon}(-S_h^{n-1}) : \left(\mathbb{C}(S_h^{n-1})\bar{\varepsilon}(-S_h^{n-1})\right)\right] \\
&= \frac{1}{2}\varepsilon(u_h^n) : \left((S_h^n - S_h^{n-1})(\mathbb{C}_2 - \mathbb{C}_1)\varepsilon(u_h^n)\right) \\
&- \left(\bar{\varepsilon}S_h^n : \mathbb{C}(S_h^n)\varepsilon(u_h^n) - \bar{\varepsilon}S_h^{n-1} : \mathbb{C}(S_h^{n-1})\varepsilon(u_h^n)\right) \\
&+ \frac{1}{2}\,\left(\bar{\varepsilon}S_h^n : (\mathbb{C}(S_h^n)\bar{\varepsilon}S_h^n) - \bar{\varepsilon}S_h^{n-1} : \left(\mathbb{C}(S_h^{n-1})\bar{\varepsilon}S_h^{n-1}\right)\right) \\
&= \frac{1}{2}\varepsilon(u_h^n) : \left((S_h^n - S_h^{n-1})(\mathbb{C}_2 - \mathbb{C}_1)\varepsilon(u_h^n)\right) \\
&- \bar{\varepsilon} : \left(\mathbb{C}_1\varepsilon(u_h^n)(S_h^n - S_h^{n-1})\right) - \bar{\varepsilon} : \left((\mathbb{C}_1 - \mathbb{C}_2)\varepsilon(u_h^n)(S_h^n + S_h^{n-1})(S_h^n - S_h^{n-1})\right) \\
&+ \frac{1}{2}\bar{\varepsilon} : \left(\mathbb{C}_1\bar{\varepsilon}\,(S_h^n + S_h^{n-1})(S_h^n - S_h^{n-1})\right) \\
&+ \frac{1}{2}\bar{\varepsilon} : \left((\mathbb{C}_2 - \mathbb{C}_1)\bar{\varepsilon}\,(S_h^n - S_h^{n-1})((S_h^n)^2 + S_h^n S_h^{n-1} + (S_h^{n-1})^2)\right)\,,
\end{aligned}$$

leading to

$$\begin{aligned}
\frac{W(S_h^n, u_h^n) - W(S_h^{n-1}, u_h^n)}{S_h^n - S_h^{n-1}} &= \left(\frac{1}{2}\varepsilon(u_h^n) - \bar{\varepsilon}(S_h^n + S_h^{n-1})\right) : ((\mathbb{C}_2 - \mathbb{C}_1)\varepsilon(u_h^n)) \\
&- \bar{\varepsilon} : (\mathbb{C}_1\varepsilon(u_h^n)) + \frac{1}{2}\bar{\varepsilon} : \left(\mathbb{C}_1\bar{\varepsilon}\,(S_h^n + S_h^{n-1})\right)
\end{aligned}$$

$$+ \quad \frac{1}{2}\bar{\varepsilon} : \left((\mathbb{C}_2 - \mathbb{C}_1)\bar{\varepsilon}\,\left((S_h^n)^2 + S_h^n S_h^{n-1} + (S_h^{n-1})^2\right)\right) .$$

We simplify the notation by writing

$$\frac{W(S_h^n, u_h^n) - W(S_h^{n-1}, u_h^n)}{S_h^n - S_h^{n-1}} = e_1\,(S_h^n)^2 + e_2\,S_h^n + e_3 \qquad (10.25)$$

with

$$\begin{aligned}
e_1 &= \frac{1}{2}\bar{\varepsilon} : ((\mathbb{C}_2 - \mathbb{C}_1)\bar{\varepsilon}) \\
e_2 &= -\bar{\varepsilon} : ((\mathbb{C}_2 - \mathbb{C}_1)\varepsilon(u_h^n)) + \frac{1}{2}\bar{\varepsilon} : (\mathbb{C}_1\bar{\varepsilon}) + \frac{1}{2}\,S_h^{n-1}\bar{\varepsilon} : ((\mathbb{C}_2 - \mathbb{C}_1)\bar{\varepsilon}) \\
e_3 &= \frac{1}{2}\left(\varepsilon(u_h^n) - \bar{\varepsilon}S_h^{n-1}\right) : ((\mathbb{C}_2 - \mathbb{C}_1)\varepsilon(u_h^n)) - \bar{\varepsilon} : (\mathbb{C}_1\varepsilon(u_h^n)) \\
&\quad + \frac{1}{2}\bar{\varepsilon}S_h^{n-1} : (\mathbb{C}_1\bar{\varepsilon}) + \frac{1}{2}\bar{\varepsilon} : \left((\mathbb{C}_2 - \mathbb{C}_1)\bar{\varepsilon}\,(S_h^{n-1})^2\right) .
\end{aligned}$$

Remark 10.1. As an alternative to the implementation of the Banach fixed-point solution, a Newton method could be applied. For future research, the implementation is provisionally outlined in the Appendix 8.

10.3 Comparison of the models

We restrict the following numerical simulations to isotropic linear elastic materials. We assume phase-dependent constant entries in the linear elasticity tensors and phase-dependent constant eigenstrains. We choose the parameters in such a way that we get clearly visible displacements and can compare both models.

Furthermore, we show another example in Chapter 10.4, comparing our results to results of published literature adapted to physical experiments.

To compare both phase field models, we apply inhomogeneous Dirichlet boundary conditions on the right-hand side for the displacement $u(x)$, so the boundary conditions read

$$\begin{aligned}
u_1(x) &= 0.0 && \forall x \in \partial\Omega_1 , \\
u_2(x) &= 0.0 && \forall x \in \partial\Omega_2 , \\
u_1(x) &= -0.03 && \forall x \in \partial\Omega_3 , \\
u_2(x) &= 0.0 && \forall x \in \partial\Omega_4 ,
\end{aligned}$$

with $u_1(x)$ denoting the horizontal displacement and $u_2(x)$ denoting the vertical displacement, see Fig. 10.2 with $u_D = u_1(x)$ on Ω_3. The initial- and boundary conditions for the order parameter remain given by Eq. (7.20) and Eq. (7.21).

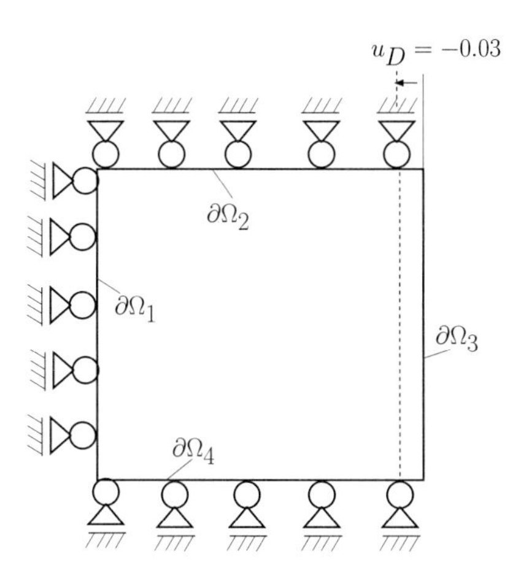

Figure 10.2: Dirichlet boundary condition.

We set the isotropic elasticity tensors to

$$\mathbb{C}_1 = \begin{pmatrix} 1.2 & 0.4 & 0.0 \\ 0.4 & 1.2 & 0,0 \\ 0.0 & 0.0 & 0.4 \end{pmatrix} 10^3 \,; \quad \mathbb{C}_2 = 2\mathbb{C}_1 \,.$$

For the implementation, we rewrite the strain tensor, defined by Eq. (8.4), based on the symmetry of $\varepsilon(u)$ and $\bar{\varepsilon}$ and the assumption of plane strain, see Appendix 5, setting the third direction of the strains to zero, to

$$\tilde{\varepsilon}(S,u) = \begin{pmatrix} \tilde{\varepsilon}_{11} & \tilde{\varepsilon}_{12} & \tilde{\varepsilon}_{13} \\ \tilde{\varepsilon}_{21} & \tilde{\varepsilon}_{22} & \tilde{\varepsilon}_{23} \\ \tilde{\varepsilon}_{13} & \tilde{\varepsilon}_{32} & \tilde{\varepsilon}_{33} \end{pmatrix} = \begin{pmatrix} \tilde{\varepsilon}_{11} & \tilde{\varepsilon}_{12} & 0 \\ \tilde{\varepsilon}_{12} & \tilde{\varepsilon}_{22} & 0 \\ 0 & 0 & 0 \end{pmatrix} \overset{\text{Implem.}}{\Longrightarrow} \begin{pmatrix} \tilde{\varepsilon}_{11} \\ \tilde{\varepsilon}_{22} \\ \gamma_{12} \end{pmatrix} . \tag{10.26}$$

Here we defined $\tilde{\gamma}_{12} = 2\bar{\varepsilon}_{12}$ and we omitted the arguments u and S for clarity. The eigenstrain vector, we implement, reads

$$\bar{\varepsilon} = \begin{pmatrix} \bar{\varepsilon}_{11} & 0 & 0 \\ 0 & \bar{\varepsilon}_{22} & 0 \\ 0 & 0 & 0 \end{pmatrix} \overset{\text{Implem.}}{\Longrightarrow} \begin{pmatrix} \bar{\varepsilon}_{11} \\ \bar{\varepsilon}_{22} \\ 0 \end{pmatrix}.$$

We define $\lambda = \mu = \nu = 0.1, dt = 0.001$ and compare solutions for $nx = 240$ for the Allen-Cahn model and $nx = 80$ for the hybrid model. Figs. 10.3, 10.4 and 10.5 show the shrinking circles and the respective displacement plots. Since we have no analytical solution for the shrinking circle, we compare the time-dependent radius development of both models which is done in Fig. 10.6.

The curves in Fig. 10.6 look quantitatively similar. We leave it with this example and show more detailed investigations in Chapter 10.4: the example of a growing martensite nucleus, adapted to experimental data.

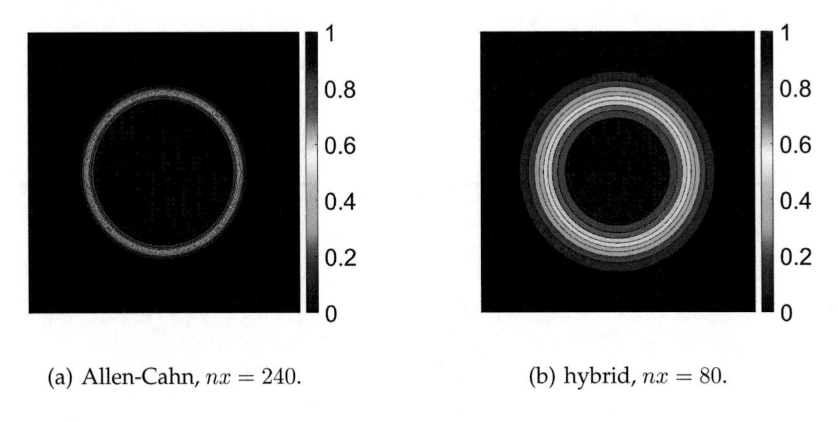

(a) Allen-Cahn, $nx = 240$.

(b) hybrid, $nx = 80$.

Figure 10.3: Comparison of the circles with $\lambda = \mu = \nu = 0.1, dt = 0.001$, $dx = 0.01$ at $t = 1.0$.

So far we can state: Because there are no classical solutions for the elastic phase field problem, we compared the numerical simulation results to some examples from [4] with a so-called ersatzstress term.

Further, we compared solutions of both models taking the Allen-Cahn model as reference model.

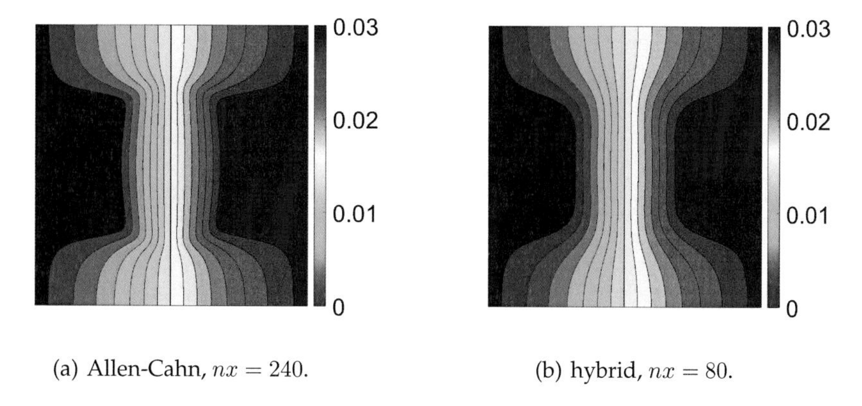

(a) Allen-Cahn, $nx = 240$. (b) hybrid, $nx = 80$.

Figure 10.4: Comparison of the displacements u_1 with $\lambda = \mu = \nu = 0.1$, $dt = 0.001, dx = 0.01$ at $t = 1.0$.

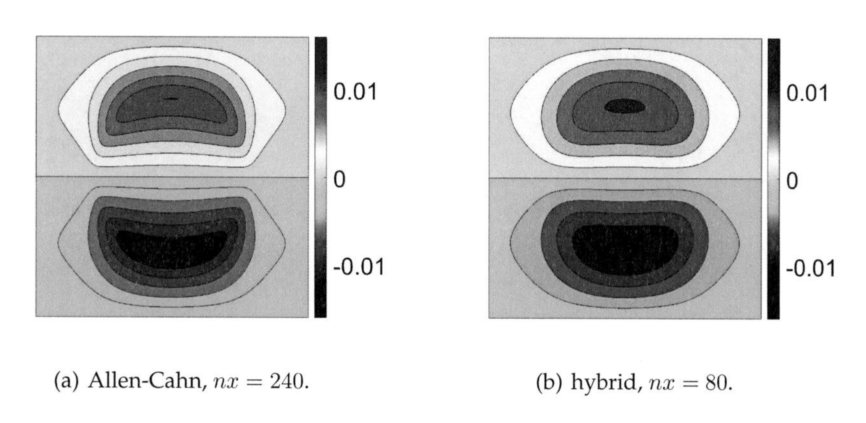

(a) Allen-Cahn, $nx = 240$. (b) hybrid, $nx = 80$.

Figure 10.5: Comparison of the displacements u_2 with $\lambda = \mu = \nu = 0.1$, $dt = 0.001, dx = 0.01$ at $t = 1.0$.

It seems that the hybrid model behaves similar to the Allen-Cahn model applied on the same configuration with adapted parameters explained above. For the phase field problem coupled to linear elasticity and given eigenstrain, the curves of the Allen-Cahn model and the hybrid model look similar with a considerable numerical advantage for the hybrid model.

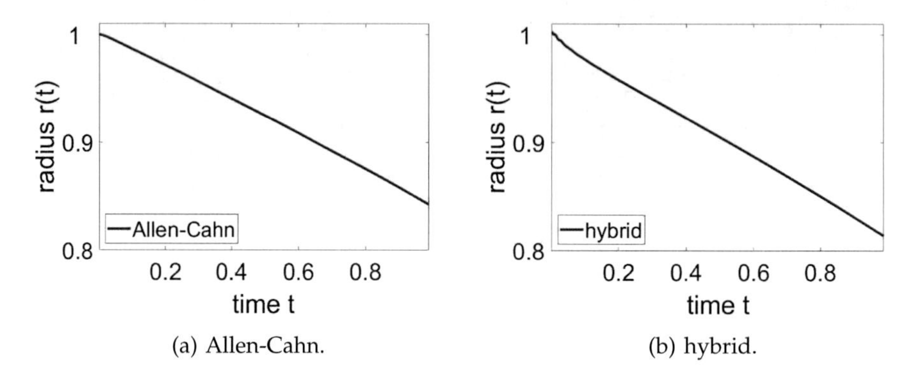

(a) Allen-Cahn.

(b) hybrid.

Figure 10.6: Decreasing radii.

The problem, identified at the end of Chapter 7.2, the radius stagnation, appears with the hybrid model at the same critical size applied to elastic problems, see Fig. 10.7.

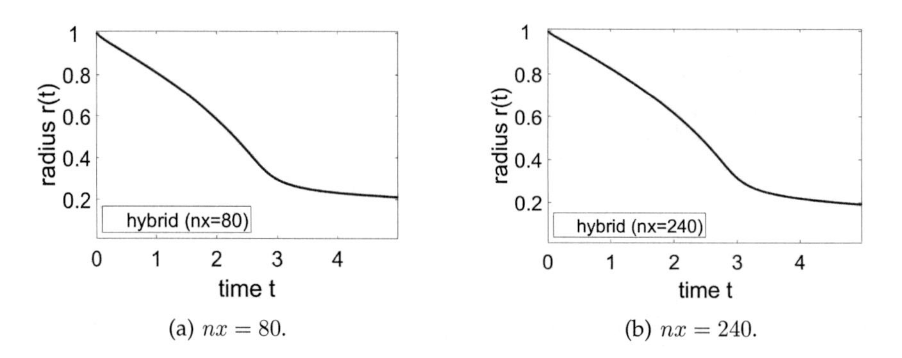

(a) $nx = 80$.

(b) $nx = 240$.

Figure 10.7: Hybrid behaviour at the critical size.

We will at last look upon a problem on very small scales with reference results verified by molecular-dynamic experiments in literature, namely the martensite transformation. For this purpose, the frame of martensite transformations used for this last example, will be explained.

10.4 A coupled elasticity problem - martensite

The transformation from austenite to martensite is of high technical interest. By a sudden undercooling the austenite steel changes its crystal structure from a metastable cubic face-centered austenite phase to a stable cubic body-centered martensite phase with different material properties, see Fig. 10.8.

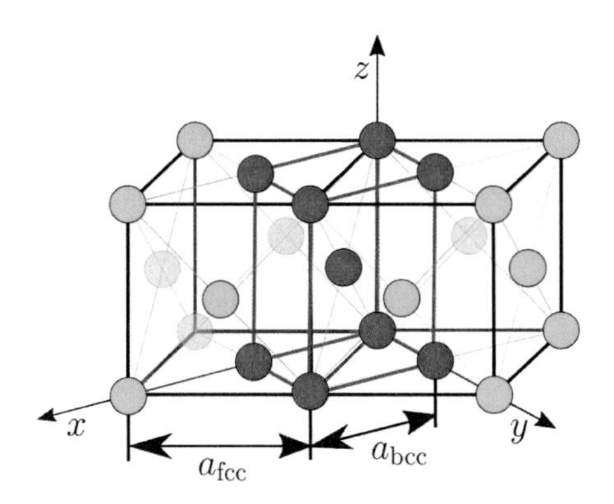

Figure 10.8: Martensite transformation, R. Müller [2015].

Since the undercooling takes place fast, the atoms are cumbered to change their places and the original topology is maintained on the costs of inner stresses and strains. The new martensite phase grows spicolar into the present austenite phase and causes a complex microstructure. Due to the eigenstrain of martensite versus the austenite domain, the atomic lattice deforms causing micro cracks and plastic deformations. Alloys of iron and carbon, copper, zinc, nickel or titanium show such a behaviour [11]. A shape memory alloy effect occurs if the martensite material is deformed plastically. As soon as the deformed martensite material is reheated, it regains its original austenite cubic face-centered structure and simultaneously turns back to its primary shape.

The generally temperature-dependent double well potential for martensite transformations determines the energy barrier that has to be overcome for the transformation process, see figure 10.9.

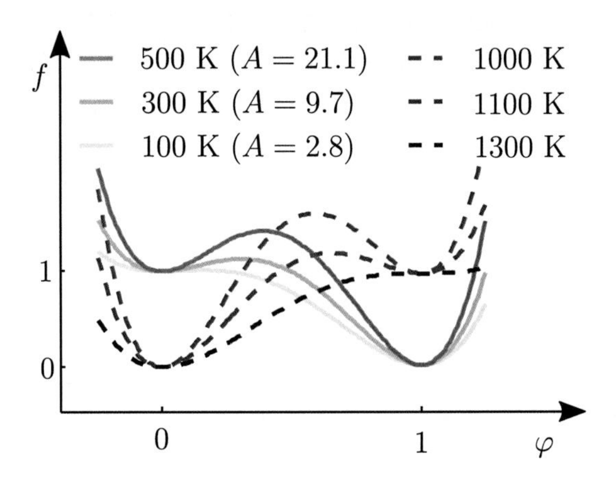

Figure 10.9: Temperature dependent double well potential martensite, R. Müller [2015].

The topology of the martensite inclusions associated to certain material properties can be influenced by specific external loads and/or material-dependent inner eigenstrains. To produce functional materials, combined temperature and loading procedures are applied to achieve direction dependent properties like hardness or Young's modulus parameter.

To verify our implementation we will examine a martensite nucleus included in an austenitic matrix under mechanical loading and compare it to some data given in literature. Since we do not address temperature dependent phase field simulations in the present work, we choose a double well potential for a constant temperature from [78] and compare our results to the results of this publication. We found this a good example for our former considerations. The interfacial energy of the included martensite nucleus is small compared to the volume forces and the relation of inner forces and interfacial energy influences our error terms discussed in Chapter 5.

The double well potential, shown in Fig. 10.10, does not fulfil all the requirements of the proofs in [4]. However, analogous proofs are very likely possible and adapting the proofs for different potentials is another project for future research.

The images of growing inclusions, we want to simulate now, are confirmed in [78] by molecular dynamic computations. The paper defines the calibration constants $\kappa_S = 1.3592$, $\kappa_G = 0.6960$, the interface energy density $G = 0.1\,J/m^2$, the interface width determining parameter $L = 5\ nm$ and the mobility factor $M = 10^{-6}\,m^3/Js$. The corresponding rescaled Allen-Cahn equation is in the notation of [78] given by

$$\dot{c_i} = -M\left[\frac{\partial W}{\partial c_i} + G\left(\frac{\kappa_s}{L}\frac{\partial f}{\partial c_i} - \kappa_G L \Delta c_i\right)\right], \quad i = 1, 2. \tag{10.27}$$

For $i = 1$ the variable c_i denotes our order parameter S and f is the asymmetric double well potential specified as

$$\psi(S) = 1 + 0.075S^2 - 4.15S^3 + 3.075S^4, \tag{10.28}$$

plotted in Fig. 10.10.

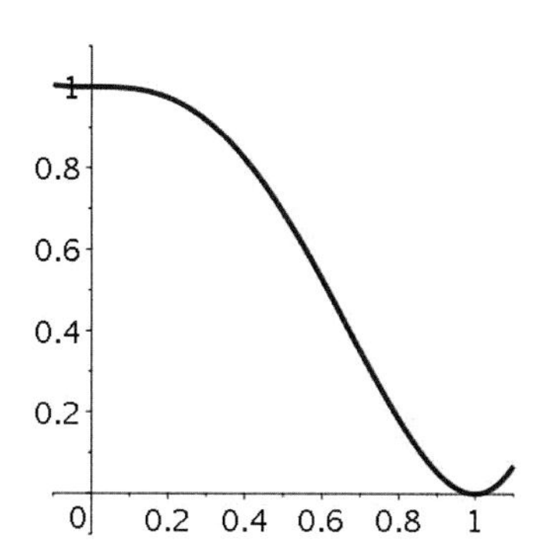

Figure 10.10: Double well potential martensite.

Compared to our previous examples, the two minima of the austenite-martensite double well potential have different values. The first local minimum represents the metastable austenite phase and the second global minimum represents the martensite phase as the stable phase. The first minimum in $S = 0$ is almost invisible in Fig. 10.10 and the transformation process will start upon very small applied forces and /or temperature changes.

In this experiment, the energy that forces the order parameter value to leave the first minimum, is provided by mechanical loadings. We speak of mechanical induced martensite transformation opposite to thermal induced martensite transformation.

The parameter G can be identified with the interface energy. Comparing the model in [78] and adapting their notation

$$\dot{S} = -M\left(W_S + G\left(\frac{\kappa_s}{L}\psi_S - \kappa_g L\Delta S\right)\right) \tag{10.29}$$

to the Allen-Cahn model

$$\dot{S} = \frac{\tilde{c}}{(\mu\lambda)^{1/2}}\left(W_S + \lambda^{1/2}\left(\frac{1}{(\mu\lambda)^{1/2}}\psi_S - (\mu\lambda)^{1/2}\Delta S\right)\right) \tag{10.30}$$

yields the system

$$M = \frac{\tilde{c}}{(\mu\lambda)^{1/2}}, \quad \frac{G\kappa_s}{L} = \frac{1}{\mu^{1/2}}, \quad G\kappa_g L = \mu^{1/2}\lambda, \tag{10.31}$$

giving us the expressions

$$\tilde{c} = ML\left(\frac{\kappa_g}{\kappa_s}\right)^{1/2}, \quad \mu = \left(\frac{L}{G\kappa_S}\right)^2, \quad \lambda = G^2\kappa_g\kappa_s\,. \tag{10.32}$$

To analyse the dimension of μ and λ we insert Eq. (10.32) into Eq. (5.14) and, replacing the index 1 by AC, we compare the interface velocities

$$s_{AC} = \frac{\tilde{c}}{c_1}\left(n\cdot[\hat{C}]n + c_1\lambda^{1/2}\kappa_\Gamma + O(\mu^{1/2})\right), \tag{10.33}$$

$$s_{Mart.} = \frac{ML}{c_1}\left(\frac{\kappa_g}{\kappa_s}\right)^{1/2}\left(n\cdot[\hat{C}]n + c_1 G\sqrt{\kappa_g\kappa_s}\kappa_\Gamma + O\left(\frac{L}{G}\right)\right). \tag{10.34}$$

This demonstrates the apparent meaning of the parameter μ for this rescaled Allen-Cahn model, because the interfacial energy parameter G is in the denominator of the last term in Eq. (10.34). Thus, a very small or negligible interfacial energy enlarges the error term in a strong way. The interface width parameter L has to become very small to counteract this effect. On the other hand, a small value of L requires a finer mesh for numerical simulations and therefore, increases the simulation time.

Comparing Eq. (10.33) and Eq. (10.34) we recognise

$$\text{err} = O(\mu^{1/2}) = O\left(\frac{L}{G}\right) = O\left(\frac{\text{interface width}}{\text{interfacial energy}}\right). \tag{10.35}$$

With the given values and Eq. (10.31) we get

$$\begin{aligned} \tilde{c} &= 3.578 \cdot 10^{-15}, \\ \mu &= 1.353 \cdot 10^{-15}, \\ \lambda &= 9.463 \cdot 10^{-3}. \end{aligned} \tag{10.36}$$

As in [78] we set the elasticity tensors to

$$\begin{aligned} \mathbb{C}_A &= \begin{pmatrix} 1.40 & 0.84 & 0.00 \\ 0.84 & 1.40 & 0.00 \\ 0.00 & 0.00 & 0.28 \end{pmatrix} \cdot 10^5 \frac{\text{N}}{\text{mm}^2}, \\ \mathbb{C}_M &= 1.1\,\mathbb{C}_A \end{aligned}$$

and the eigenstrain to

$$\bar{\varepsilon} = \begin{pmatrix} -0.1 & 0.1 & 0.0 \end{pmatrix}^{\mathsf{T}}.$$

In the numerical examples in Chapter 7 and Chapter 10 we chose parameters and boundary conditions in such a way that the contexts, we studied, became clear. Now we want to compare our simulations to results of a publication that was based on molecular-dynamical computations. The goal is to show that the hybrid model provides comparable results also in real physical situations.

In the following, we show simulations comparing the Allen-Cahn model to the hybrid model in the context of martensite transformations. We select a time step size of $dt = 0.001$ and discretise the domain with 400×400 nodes.

Fig. 10.11 shows the simulation result with the Allen-Cahn model matching the picture from [78].

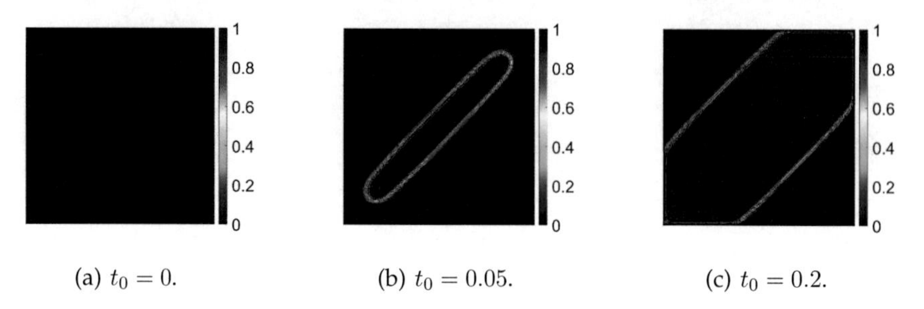

(a) $t_0 = 0$. (b) $t_0 = 0.05$. (c) $t_0 = 0.2$.

Figure 10.11: Growth of martensite phase with Allen-Cahn.

To simulate the problem with the hybrid model we have to set $\nu = \lambda$, see Table 5.2.

Remark 10.2. Strictly speaking, c_1 in Eq. (10.33) does not correspond exactly to ω_1 in Eq. (3.18), as permissibly assumed for the non-elastic case, if we consider elasticity as in this martensite example. For the linear elastic case instead of the double-well potential Eq. (2.5) determining the value of c_1 by Eq. (2.17), the effective double-well potential from [4] would have to be used to calculate $\omega_1 \neq c_1$. This effective potential contains an additional expression depending on the eigenstrain $\bar{\varepsilon}$ and the elasticity tensor $\mathbb{C}(S)$. A rough estimation, not carried out here, however showed that this term is not significantly influencing the value of ω_1 in this example, so we continue with the assumption that $c_1 \approx \omega_1$.

With $\nu = \lambda$ and Eq. (5.30) we calculate a relationship of $re_B = 0.0367$, giving an about 27 times wider interface zone for the hybrid model. It is obvious that we receive a huge interface width and thus the interface reaches the centerline of the ellipse immediately, destroying the simulation, see Fig. 10.12. This conclusion is a result of our investigations into the conditions of the critical size in Chapter 7.2 and helps us here to better understand these simulations.

To achieve $re_A = 1$, we set $\nu = \lambda\mu$. This means that we reduce the influence of the interfacial energy, which we can do, because martensite transformations have very little interfacial energy. As explained before we regard the $O()$-terms in Eq. (5.25) and Eq. (5.26) as error terms beyond the elastic driving force.

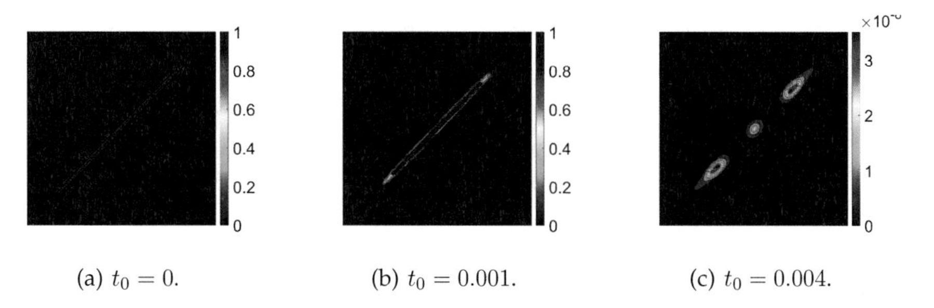

(a) $t_0 = 0$. (b) $t_0 = 0.001$. (c) $t_0 = 0.004$.

Figure 10.12: Growth of martensite phase with hybrid.

Beneath adjusting the width of the interface we expect a better convergence by minimising the error between the classical and the numerical solution as explained in Chapter 5.4.

Anyhow, the result is not satisfactory, because the interface of the hybrid model does not move at all, see Fig. 10.13. Neither mesh refinement nor time refinement seems to change this fact.

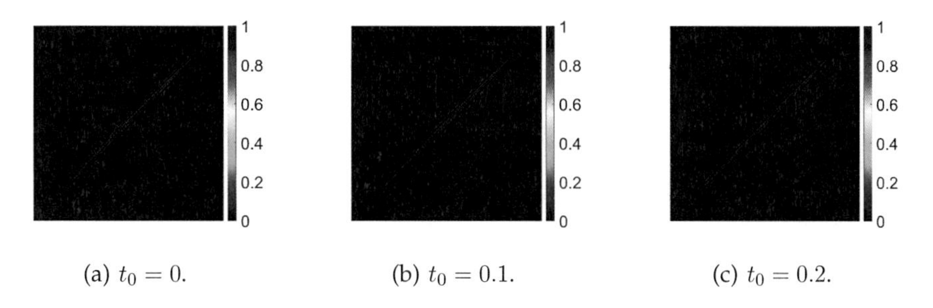

(a) $t_0 = 0$. (b) $t_0 = 0.1$. (c) $t_0 = 0.2$.

Figure 10.13: Growth of martensite phase with hybrid and $\nu = \mu\lambda$.

Trying to find out which mechanism initiates the interface to move, we vary the function f_2, explained in Table 5.2, and recognise that increasing the mobility constant causes the nucleus to grow, see Fig. 10.14. We notice that we get results comparable to the Allen-Cahn simulation in Fig. 10.11 by setting $f_2 = 1.0e8 \cdot \tilde{c}/c_1$, see Fig. 10.15.

The question to be answered is obvious. Before we said that setting $f_2 = \tilde{c}/c_1$ leads to comparable results for the Allen-Cahn model and the hybrid model. So we have to explain the very different choice of $f_2 = 1.0e8 \cdot \tilde{c}/c_1$ for a comparable simulation in this situation.

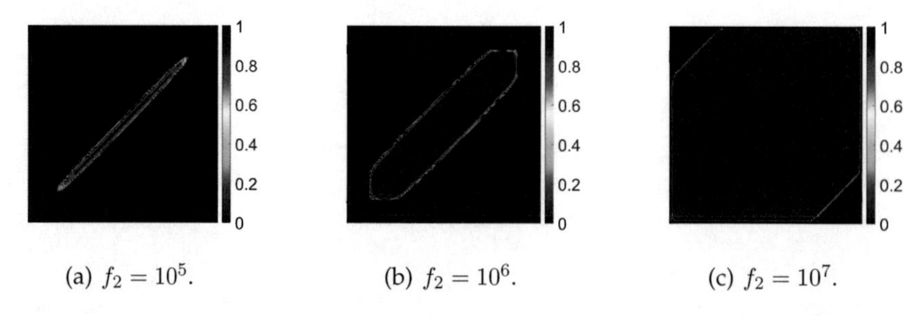

(a) $f_2 = 10^5$. (b) $f_2 = 10^6$. (c) $f_2 = 10^7$.

Figure 10.14: Growth of martensite phase with hybrid and $\nu = \mu\lambda$ at $t_1 = 5$.

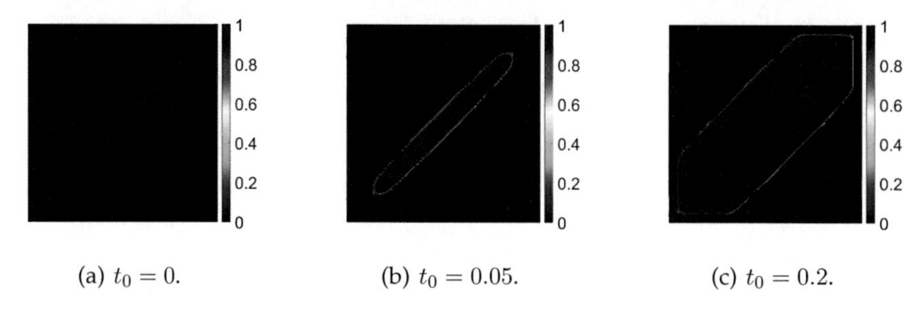

(a) $t_0 = 0$. (b) $t_0 = 0.05$. (c) $t_0 = 0.2$.

Figure 10.15: Growth of martensite phase with hybrid $f_2 = 10^8 \cdot \tilde{c}, \nu = \mu\lambda$.

To consider this, we have a look at Eq. (5.25) and Eq. (5.26), where we have an error term of $O(\lambda^{1/2}) + O(\mu^{1/2})$ for the Allen-Cahn model and an error term of $O(\nu^{1/2})$ for the hybrid model. Additionally, we have to transfer $O(\nu)$ to $O(\mu\lambda)$ for our last examples. So with the values of Eq. (10.36) we compare errors of $O(10^{-1})$ to errors of approximately $O(10^{-8})$.

If we try to explain the choice of $f_2 = 10^8 \cdot \tilde{c}/c_1$ for the results of the last simulation in Fig. 10.15 this way, this would mean that the solution of the hybrid model is very much closer to the exact sharp phase field solution. This in turn would mean that the parameters chosen for the Allen-Cahn model contain a large error regarding the sharp interface solution.

Anyhow, the parameters given in [78] were adapted by physical experiments and not by the best possible approach to the sharp interface solution. So it should also be valid to choose the parameters for the hybrid model in the best way to imitate the behaviour of the given nucleus.

It might be interesting for future research to examine both models in terms of such problems including very small scales.

10.5 Conclusion

In the present work, we studied the hybrid model as a new phase field model compared to the well-known Allen-Cahn model. We examined the phase field equations in terms of well-posedness, convergence behaviour, numerical errors and thermodynamic consistency. A semi-discrete approximation was introduced and transformed into a fully-discrete numerical scheme. The numerical Finite Element implementation was applied to simple examples with existing analytical solutions. The results were compared especially regarding the convergence behaviour focusing on the scaling of the interface widths.

The model was expanded to a coupled elasticity term connected to a constitutive equation and a momentum balance. Proofs for well-posedness and energy decay properties of the fully discretised form were partially given and numerical examples with reference to results published in literature were simulated. The hybrid model was confirmed as a good alternative to the Allen-Cahn model, at least for cases of small interface energy. This in turn was related to the combination of error terms regarding the Allen-Cahn model we examined in detail. We argued that for vanishing interfacial energies under the condition of small error terms, the use of the Allen-Cahn model required a very fine mesh compared to the hybrid model.

Finally, our analysis showed that for simulations near singularities, like a disappearing circle or developing inclusions, the hybrid model behaves differently. In order to use the hybrid model for martensite transformations or other small-scale problems, the situation on small scales had to be given special attention. This was done by the adaption of the asymptotic parameter ν and the adaption of the mobility function.

In summary, we can say that the use of the hybrid model in certain contexts indicates a precise analysis in terms of interface energy, simulation effort, domain size and other physical factors. For special situations with low interfacial energy, adapting the parameters on very small scales, the hybrid model can describe phase field problems with an essential advantage. It is certainly worth continuing the research on the hybrid model. Especially the numerical implementation for small inclusions, different boundary conditions, more-phase problems, different material laws, among others, deserve further investigations to make the hybrid model widely applicable.

A APPENDIX

1 Proof in [4]

We consider the interface neighbourhood in the notation of [4], defined by Eq. (3.17), and the indicator function $\phi(x,t) \in C^\infty$ defining a point inside ($\phi = 1$) or outside ($\phi = 0$) of $\mathcal{U}$. The function $v(x,t)$ in the equation system

$$\begin{aligned} u^{(\nu)}(x,t) &= \phi(x,t) \sum_{i=0}^{1} \nu^{\frac{1+i}{2}} u_i(\eta, \zeta, t) + v(x,t), \\ S^{(\nu)}(x,t) &= \phi(x,t) \sum_{i=0}^{1} \nu^{\frac{i}{2}} S_i(\eta, \zeta, t) + (1 - \phi(x,t)) \hat{S}(x,t), \\ T^{(\nu)}(x,t) &= D\left(\varepsilon\left(\nabla u^{(\nu)}(x,t)\right) - \bar{\varepsilon} S(\nu)(x,t)\right), \end{aligned}$$

stands for the additional displacement inside and the only displacement outside of $\mathcal{U}$ and $\hat{S}(x,t)$ denotes the value of the order parameter outside of the Γ-neighbourhood. These approaches are inserted into the system (3.10) - (3.12) in order to determine the unknowns u_0, u_1, S_0 and S_1. Combining the terms of same order of ν leads to a system of second order differential equations

$$T_0'(\zeta)n = 0, \tag{A.1.1}$$

$$T_1'(\zeta)n = -\mathrm{div}_\eta T_0(\zeta), \tag{A.1.2}$$

$$\tilde{\psi}_S(S_0(\zeta)) - S_0''(\zeta) = 0, \tag{A.1.3}$$

$$\tilde{\psi}_{SS}(S_0(\zeta))S_1(\zeta) - S_1''(\zeta) = g_1(t,\eta,\zeta) + \omega(t,\eta), \tag{A.1.4}$$

supplemented by boundary conditions and a free energy function

$$\tilde{\psi}(t,\eta,S) = \hat{\psi}(S) - \hat{\psi}(0)(1-S) - \hat{\psi}(1)S + \frac{1}{2}p(n(t,\eta))S(1-S).$$

The local coordinate variable

$$\zeta = \frac{\xi}{\nu^{1/2}}$$

is introduced and derivates $(\cdot)'$ apply on it. Further we have the definition

$$p(n) = -\bar{\varepsilon} : D(I - P_n)\bar{\varepsilon}$$

with P_n being the projection onto S_n^3. The definitions of the right-hand side in Eq. (A.1.4) are given in [4] and depend on stresses and strains, as well as on the unknowns of the system (3.10) - (3.12).

Instead of solving Eq. (A.1.3), the initial value problem

$$\partial_\zeta S_0(n,\zeta) = \sqrt{2\tilde{\psi}(n, S_0(n,\zeta))}, \quad S_0(n,0) = \frac{1}{2}$$

is examined and yields $$S_0'' = \frac{\tilde{\psi}_S(n, S_0(n,\zeta)) \overbrace{\partial_\zeta S_0}^{\sqrt{2\tilde{\psi}(n,S_0(n,\zeta))}}}{\sqrt{2\tilde{\psi}(n, S_0(n,\zeta))}}.$$

The existence of a unique bounded and strictly increasing solution $S_0(n,\zeta)$ is shown in the proof of Theorem 2.3 in [4]. To show existence of solutions, Eq. (A.1.4) is written in terms of

$$LS_1 = F_1 \quad \text{with} \quad L = \tilde{\psi}_{SS}(S_0(\zeta)) - \partial_{\zeta\zeta}\,.$$

Deriving Eq. (A.1.3) with respect to ζ yields

$$(\tilde{\psi}_{SS} - \partial_{\zeta\zeta})S_0' = 0\,,$$

so that $S_0' = \partial_\zeta S_0$ is an eigenfunction to the eigenvalue zero of the linear boundary value problem

$$\tilde{\psi}_{SS}(S_0(\zeta))S_1(\zeta) - S_1''(\zeta) = F_1(t,\eta,\zeta)\,.$$

The spectral theory of self-adjoint differential operators states that

$$(LS_1, S_0')_\Omega = (S_1, LS_0')_\Omega = 0$$

and therefore the boundary value problem (A.1.4) has solutions if

$$\int_{a(t,\eta)}^{b(t,\eta)} F_1(t,\eta,\zeta)S_0'(t,\eta,\zeta)\, d\zeta = 0$$

holds; see [4], theorem 2.4 and the respective proof. Extending the functions S_0 and S_1 to $\zeta \in \mathbb{R}\setminus[a(t,\eta), b(t,\eta)]$ with

$$S_0(t,\eta,\zeta) = \begin{cases} 0, & \zeta < a(t,\eta), \\ 1, & \zeta > b(t,\eta), \end{cases} \quad; \qquad S_1(t,\eta,\zeta) = 0\,,$$

the authors show that the Eqs. (A.1.1) and (A.1.2), supplemented by the boundary conditions, can be solved with unique continuously differentiable solutions for the functions of (3.13) - (3.15). Details of the one-dimensional proofs and the associated assumptions, e.g. symmetry conditions applied on the double well potential, can be studied in the original literature. For unsymmetric double well potentials the proofs can be adjusted.

Inserting the asymptotic solutions $(u^{(\nu)}), T^{(\nu)}), S^{(\nu)})$ into (3.10) - (3.12), the authors show the estimates

$$\left|\operatorname{div}_x T^{(\nu)}(x,t) + b(x,t)\right| \leq \begin{cases} K_1 \nu^{1/2}, & (x,t) \in \Gamma[\nu], \\ K_2 \nu, & (x,t) \in Q \backslash \Gamma[\nu], \end{cases}$$

$$||\partial_t S^{(\nu)} \quad + f(\psi_S\left(S^{(\nu)}, \varepsilon(\nabla_x u^{(\nu)})\right) - \nu \Delta_x S^{(\nu)} |\nabla_x S^{(\nu)}|||_{L^\infty(V)}$$

$$\leq \begin{cases} K_3 \nu^{1/2}, & \text{for } V = \Gamma_k[\nu], \\ K_4, & \text{for } V = \tilde{\Gamma}_k[\nu], \\ 0, & \text{for } V = Q \backslash \Gamma_k[\nu]. \end{cases} \tag{A.1.5}$$

The regions $\Gamma_k[\nu]$ and $\tilde{\Gamma}_k[\nu]$ are attached to $\Gamma(t)$ and $Q \in \mathbb{R}^3$ denotes the whole domain. For a two phase-system, described by an order parameter $S \in [0,1]$, this order parameter goes within the neighbourhood domains smoothly and continuous from 0 to 1. The outer expansions are $S(x,t) = 0$ for the one phase and $S(x,t) = 1$ for the other.

With the proof of the estimate Eq. (A.1.5) in [4], the normal interface speed s appears with

$$s = -\partial_t \xi \, .$$

2 Four-node two-dimensional Finite Elements

To connect our global system with the local configuration, we divide our domain $\bar{\Omega}$ in standard finite elements; see Fig. A.2.1, and give them local and global node numbers; see Fig. A.2.2.

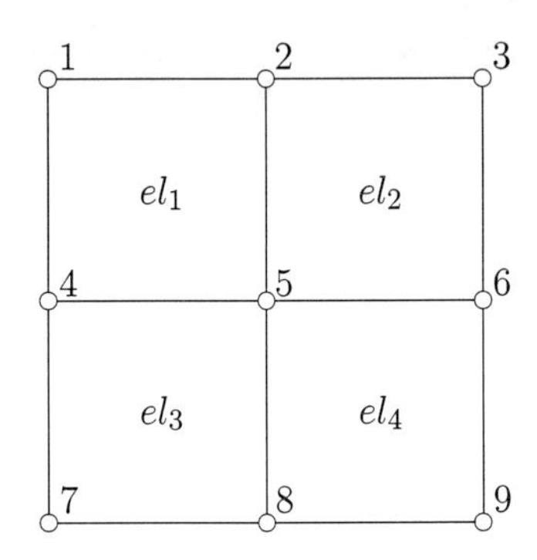

Figure A.2.1: Domain with four Finite Elements

Following the notation from [55], we specify the four-noded standard element $\Omega_{st} = [(0,1) \times (0,1)]$ with the local coordinates $\boldsymbol{\xi} = \{\xi, \eta\}^{\mathsf{T}}$. Here we define the standard bilinear shape functions

$$\begin{aligned}
N_1(\xi,\eta) &= (1-\xi)(1-\eta)\,, \\
N_2(\xi,\eta) &= \xi(1-\eta)\,, \\
N_3(\xi,\eta) &= \xi\eta\,, \\
N_4(\xi,\eta) &= (1-\xi)\eta\,, \\
& \qquad\qquad \text{for} \quad \xi,\eta \in [0,1]\,.
\end{aligned} \tag{A.2.1}$$

We choose the same shape functions for the unknowns and the test functions in Ω_{st} expressed by

$$v_{S,loc,h}^{(el)} = \sum_{i=1}^{4} N_i(\xi,\eta) v_{Si}\,; \qquad S_{loc,h}^{(el)} = \sum_{i=1}^{4} N_i(\xi,\eta) \hat{S}_i\,. \tag{A.2.2}$$

To transfer these terms into the global matrices, they have to be mapped first from the local element reference configuration Ω_{st} to the global element reference configuration Ω_{el}, which represents the actual shape of the respective element. This is done by a mapping

$$\mathbf{X} = \{x, y\}^{\mathsf{T}} = \mathbf{\Phi}^{e}(\xi, \eta) \tag{A.2.3}$$

with $\mathbf{X}$ standing for the coordinates (x, y) of Ω_{el} and $\mathbf{\Phi}$ for the mapping applied on functions in (ξ, η) in Ω_{st}. For further details see [55].

The entry of the terms, calculated by the Gaussian quadrature in Ω_{st} and then projected onto Ω_{el}, into the global element matrices is done by assignment rules, in which the node numbers of the element configuration Ω_{el} are assigned to node numbers of the global configuration $\bar{\Omega}$. We recognize this assignment in Fig. A.2.2.

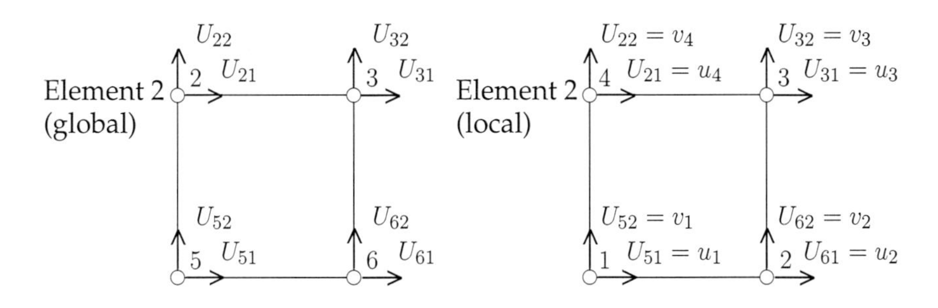

Figure A.2.2: Global and local coordinates and unknowns

The unknowns $U_{ij}, i = 1, N;\ j = 1, 2$ and $u_i,\ v_i,\ i = 1, 4$ correspond to the unknown displacements u_h^n in Chapter 10.2. The scalar unknown order parameter values S_h^n in Chapter 7.1 are assigned in a similar way not shown in Fig. A.2.2. For the assembling scheme see Appendix 3.

3 Assembling

The entries of the global Finite Element matrices have values only in the fields, which belong to the respective elements and are zero everywhere else. The sizes of the element matrices calculated at the global element reference configuration Ω_{el} are of dimension $nel \times nel$, with nel for the number of nodes per element and the sizes of the global system matrices are $N \times N$ with N for the number of all nodes of the global structure.

We call the indices for the local element nodes $i_{node,loc}$, $j_{node,loc}$ and the indices for the global element nodes $I_{node,glob}$, $J_{node,glob}$. The local indices are assigned to the respective global node indices of the respective elements by a surjective mapping

$$
\begin{aligned}
i_{node,loc} &\rightarrow I_{node,glob} \\
j_{node,loc} &\rightarrow J_{node,glob} \qquad i,j = 1,4; \quad I,J = 1,N
\end{aligned} \tag{A.3.1}
$$

E.g.: if an element has the global node indices $5, 6, 3, 2$; see Fig. A.2.2, then the positions of the local element stiffness matrix $K^{(el)}$ inside the global system stiffness matrix K are assigned by

$$
\begin{aligned}
(1,1),(1,2),(1,3),(1,4) &\rightarrow (5,5),(5,6),(5,3),(5,2)\,, \\
(2,1),(2,2),(2,3),(2,4) &\rightarrow (6,5),(6,6),(6,3),(6,2)\,, \\
(3,1),(3,2),(3,3),(3,4) &\rightarrow (3,5),(3,6),(3,3),(3,2)\,, \\
(4,1),(4,2),(4,3),(4,4) &\rightarrow (2,5),(2,6),(2,3),(2,2)\,.
\end{aligned} \tag{A.3.2}
$$

4 Spatial discretisation - Finite Element method

4.1 General model without elasticity

The spatial discretisation is based on the definitions

$$\begin{aligned} S_{h,t} &:= \textstyle\sum_{I=1}^{N} N_I \hat{S}_{t_I}\,; & \nabla S_h &:= \sum_{I=1}^{N} B_I \hat{S}_I\,, \\ \nabla v_{hS} &:= \textstyle\sum_{I=1}^{N} B_I v_{SI}\,; & \psi'(S_h) &:= \sum_{I=1}^{N} N_I \psi'(\hat{S}_I) \end{aligned} \tag{A.4.1}$$

with

$$B_I = \begin{pmatrix} N_{I,x} \\ N_{I,y} \end{pmatrix}$$

for the spatial derivatives of the basis functions (A.2.1). The hat-symbol denotes the unknown nodal variables. Insertion of Eq. (A.4.1) into Eq. (6.1), using the Einstein notation, leads to

$$\int_\Omega h(|B_I\hat{S}_I|) N_I \hat{S}_{t_I} N_J v_{SJ}\, dx + \alpha \int_\Omega B_I \hat{S}_I B_J v_{SJ}\, dx = -\beta \int_\Omega N_I \psi'(\hat{S}_I) N_J v_{SJ}\, dx\,.$$

The choice of arbitrary test functions v_{SJ} yields

$$\int_\Omega h(|B_I\hat{S}_I|) N_I N_J \hat{S}_{t_I}\, dx + \alpha \int_\Omega B_I^{\mathsf{T}} B_J \hat{S}_I\, dx = -\beta \int_\Omega N_I N_J \psi'(\hat{S}_I)\, dx\,,\ I,J = 1,N\,.$$

The nodal unknowns $\hat{S}_I$ are independent of the integration domain, so we have

$$\int_\Omega h(|B_I\hat{S}_I|) N_I N_J\, dx\, \hat{S}_{t_I} + \alpha \int_\Omega B_I^{\mathsf{T}} B_J\, dx\, \hat{S}_I = -\beta \int_\Omega N_I N_J\, dx\, \psi'(\hat{S}_I)\,. \tag{A.4.2}$$

As explained in Chapter 5.1 we have $h(|\nabla_h S|) = (\mu\lambda)^{1/2}/c_1$ for the Allen-Cahn model and $h(|\nabla_h S|) = \left((N_{I,x}\hat{S}_I)^2 + (N_{I,y}\hat{S}_I)^2\right)^{-1/2}$ for the hybrid model.

The function $\psi'(\hat{S}_I)$ is calculated deriving the double well potential (2.5) only for the starting value of the implementation scheme explained in Table 7.5. The further steps use the difference quotient, evaluated by Eq. (7.15), for the non-elastic case. Thus, within the implementation the function $\psi'(\hat{S}_I)$ is substituted by a polynom $p(a_i(\hat{S}_I)^i), i = 0,3$.

With Eq. (A.4.2) we formulate

$$M_{2_{AC}} \hat{S}_t + \alpha K \hat{S} = -\beta M_1 F(\hat{S}) \tag{A.4.3}$$

for the Allen-Cahn equation with

$$M_{2_{AC}} = \frac{(\mu\lambda)^{1/2}}{c_1} \int_\Omega N_I N_J \, dx \in \mathbb{R}^{N\times N}, \quad M_1 = \int_\Omega N_I N_J \, dx \in \mathbb{R}^{N\times N}, \tag{A.4.4}$$
$$K = \int_\Omega B_I^\mathsf{T} B_J \, dx \in \mathbb{R}^{N\times N},$$
$$F = \begin{pmatrix} p(a_i(\hat{S}_1)^i) & p(a_i(\hat{S}_2)^i) & \dots & p(a_i(\hat{S}_N)^i) \end{pmatrix}^\mathsf{T} \in \mathbb{R}^N,$$
$$\hat{S}_t = \begin{pmatrix} \hat{S}_{1,t} & \hat{S}_{2,t} & \dots & \hat{S}_{N,t} \end{pmatrix}^\mathsf{T} \in \mathbb{R}^N, \quad \hat{S} = \begin{pmatrix} \hat{S}_1 & \hat{S}_2 & \dots & \hat{S}_N \end{pmatrix}^\mathsf{T} \in \mathbb{R}^N,$$

for $I, J = 1, N$ and N equal the total number of global nodes.

The integrals are calculated by computing the terms with Gaussian quadrature on the local element reference configuration Ω_{st}. Before they are assembled into the global matrices $M_1, M_{2_{AC}}, K$ and F, they are transferred to the global element reference configuration Ω_{el} by the mapping Eq. (A.2.3); see Appendix 2 and Appendix 3.

For the hybrid model the calculation goes straightforward using the same steps with

$$M_{2_h} = \frac{1}{|\nabla \hat{S}|} \int_\Omega N_I N_J \, dx \in \mathbb{R}^{N\times N} \tag{A.4.5}$$

and

$$|\nabla \hat{S}| = \left| \begin{pmatrix} N_{I,x} \hat{S}_I \\ N_{I,y} \hat{S}_I \end{pmatrix} \right|. \tag{A.4.6}$$

So we have instead of Eq. (A.4.3)

$$M_{2_h} \hat{S}_t + \alpha K \hat{S} = -\beta M_1 F(\hat{S}). \tag{A.4.7}$$

4.2 General model with elasticity

Inserting Eq. (9.11) into Eq. (9.9), discretising the arising equation and using Eq. (7.10) and the definitions

$$v_{hu} := \sum_{I=1}^{N} N_I^u v_{uI}\,, \quad \nabla v_{hu} := \sum_{I=1}^{N} B_I^u v_{uI}\,, \; B_I^u = \begin{pmatrix} N_{I,x}^u & 0 \\ 0 & N_I^u,y \\ N_{I,y}^u & N_{I,x}^u \end{pmatrix}, \tag{A.4.8}$$

we have

$$\int_\Omega \left(\mathbb{C}_1 + N_I\hat{S}_I(\mathbb{C}_2 - \mathbb{C}_1)\right)\left(B_I^u\hat{u}_I - N_I\hat{S}_I\bar{\varepsilon}\right) B_J^u v_{uJ}\, dx = \int_\Omega bN_J^u v_{uJ}\, dx\,,$$

rewritten as

$$\int_\Omega B_J^{u\mathsf{T}}\mathbb{C}(N_I\hat{S}_I)B_I^u\hat{u}_I v_{uJ}\, dx = \int_\Omega (bN_J^u + B_J^{u\mathsf{T}}\mathbb{C}(N_I\hat{S}_I)N_I\hat{S}_I\bar{\varepsilon})v_{uJ}\, dx\,. \tag{A.4.9}$$

For arbitrary test functions v_{uJ} and the unknowns $\hat{u}$ independent of the integrals, this yields the equation

$$\int_\Omega B_J^{u\mathsf{T}}\mathbb{C}(N_I\hat{S}_I)B_I^u\, dx\, \hat{u}_I = \int_\Omega bN_J^u + B_J^{u\mathsf{T}}\mathbb{C}(N_I\hat{S}_I)N_I\hat{S}_I\bar{\varepsilon}\, dx\,, \tag{A.4.10}$$

which we can write as

$$\bar{K}\hat{u} = \hat{f} \qquad \in \mathbb{R}^{2N} \tag{A.4.11}$$

with

$$\bar{K} = \int_\Omega B_J^{u\mathsf{T}}\mathbb{C}(N_I\hat{S}_I)B_I^u\, dx \quad \in \mathbb{R}^{2N\times 2N}\,, \tag{A.4.12}$$

$$\hat{u} = \begin{pmatrix} \hat{u}_1 & \hat{u}_2 & \dots & \hat{u}_N \end{pmatrix}^\mathsf{T} \in \mathbb{R}^{2N}\,, \quad \hat{f} = \int_\Omega bN_J^u + B_J^{u\mathsf{T}}\mathbb{C}(N_I\hat{S}_I)N_I\hat{S}_I\bar{\varepsilon}\, dx \in \mathbb{R}^{2N}\,.$$

The calculation of the integrals is explained at the end of Appendix 4.1.

5 Linear elasticity and Voigt Notation

In index notation, the linear elasticity relation between the strains and the stresses is given by

$$\sigma_{ij} = C_{ijkl}\varepsilon_{kl}\,. \tag{A.5.1}$$

We denote the 81 components of the elasticity tensor by C_{ijkl}, the components of the stress tensor by σ_{ij} and the components of the strain tensor by ε_{kl}. With the symmetry of the stress tensor σ_{ij}, the elasticity tensor depends on only 36 independent components and the stress tensor contracts to six independent components; see [52]. The symmetric strain tensor is given by

$$\varepsilon = \frac{1}{2}(\nabla u + \nabla u^{\mathsf{T}}) = \begin{pmatrix} \varepsilon_{11} & \varepsilon_{12} & \varepsilon_{13} \\ \varepsilon_{21} & \varepsilon_{22} & \varepsilon_{23} \\ \varepsilon_{13} & \varepsilon_{32} & \varepsilon_{33} \end{pmatrix}. \tag{A.5.2}$$

In this case, the Voigt notation assigns its components along the lines of the stress tensor in the form (11)→ (11), (22)→ (22), (33)→ (33), (12)=(21)→ (12), (23)=(32)→ (23) and (13)=(31)→ (13). We further set $\gamma_{ij} := 2\varepsilon_{ij}$.

The constitutive equation of linear elasticity for an isotropic material is defined as the relation

$$\begin{pmatrix} \sigma_{11} \\ \sigma_{22} \\ \sigma_{33} \\ \sigma_{12} \\ \sigma_{23} \\ \sigma_{13} \end{pmatrix} = \begin{pmatrix} \tilde{\lambda}+2\tilde{\mu} & \tilde{\lambda} & \tilde{\lambda} & 0 & 0 & 0 \\ \tilde{\lambda} & \tilde{\lambda}+2\tilde{\mu} & \tilde{\lambda} & 0 & 0 & 0 \\ \tilde{\lambda} & \tilde{\lambda} & \tilde{\lambda}+2\tilde{\mu} & 0 & 0 & 0 \\ 0 & 0 & 0 & \tilde{\mu} & 0 & 0 \\ 0 & 0 & 0 & 0 & \tilde{\mu} & 0 \\ 0 & 0 & 0 & 0 & 0 & \tilde{\mu} \end{pmatrix} \begin{pmatrix} \varepsilon_{11} \\ \varepsilon_{22} \\ \varepsilon_{33} \\ \gamma_{12} \\ \gamma_{23} \\ \gamma_{13} \end{pmatrix} \tag{A.5.3}$$

with the elastic constants $\tilde{\lambda}$ and $\tilde{\mu}$ calculated by

$$\tilde{\lambda} = \frac{\tilde{\nu}E}{(1+\tilde{\nu})(1-2\tilde{\nu})}, \quad \tilde{\mu} = \frac{E}{2(1+\tilde{\nu})} \tag{A.5.4}$$

from the Young's modulus E and the Poisson's ratio $\tilde{\nu}$.

Plane strain is obtained by constraining one direction (e.g. direction (33)) in a three-dimensional body, so no displacement occurs in this direction. This situations puts the components γ_{13}, γ_{23} and ε_{33} to zero.

In thin plates with forces only in plate-direction, we assume plane stress with σ_{33}, σ_{23} and σ_{13} set to zero.

Table A.5.1 lists the relations for plane stress and plane strain, derived in [53].

Table A.5.1: Reduced equation for plane stress and plane strain

case	reduced Eq.
plane stress	$\begin{pmatrix} \sigma_{11} \\ \sigma_{22} \\ \sigma_{12} \end{pmatrix} = \begin{pmatrix} \tilde{\lambda}+2\tilde{\mu} & \tilde{\lambda} & 0 \\ \tilde{\lambda} & \tilde{\lambda}+2\tilde{\mu} & 0 \\ 0 & 0 & \tilde{\mu} \end{pmatrix} \begin{pmatrix} \varepsilon_{11} \\ \varepsilon_{22} \\ \gamma_{12} \end{pmatrix}$ $\varepsilon_{11} = \frac{1}{E'}(\sigma_{11} - \tilde{\nu}\sigma_{22})$ $\varepsilon_{22} = \frac{1}{E'}(\sigma_{22} - \tilde{\nu}\sigma_{11})$ $\gamma_{12} = \frac{2(1+\tilde{\nu})}{E'}\sigma_{12}$ additional Eq.: $\sigma_{33} = \tilde{\nu}(\sigma_{11} + \sigma_{22})$ parameters: $E' = \frac{E}{(1-\tilde{\nu}^2)}, \tilde{\nu}' = \frac{\tilde{\nu}}{1-\tilde{\nu}}$
plane strain	$\begin{pmatrix} \sigma_{11} \\ \sigma_{22} \\ \sigma_{12} \end{pmatrix} = \begin{pmatrix} \tilde{\lambda}+2\tilde{\mu} & \tilde{\lambda} & 0 \\ \tilde{\lambda} & \tilde{\lambda}+2\tilde{\mu} & 0 \\ 0 & 0 & \tilde{\mu} \end{pmatrix} \begin{pmatrix} \varepsilon_{11} \\ \varepsilon_{22} \\ \gamma_{12} \end{pmatrix}$ $\varepsilon_{11} = \frac{1}{E}(\sigma_{11} - \tilde{\nu}\sigma_{22})$ $\varepsilon_{22} = \frac{1}{E}(\sigma_{22} - \tilde{\nu}\sigma_{11})$ $\gamma_{12} = \frac{2(1+\tilde{\nu})}{E}\sigma_{12}$ additional Eq.: $\varepsilon_{33} = -\tilde{\nu}(\sigma_{11} + \sigma_{22})/E$

6 Dissipation inequality

The second law of thermodynamics combined with the entropy-balance reads

$$\rho_0 \dot{s} + \operatorname{div}(\Phi) = \xi \geq 0,$$

with the density $\rho_0 \in \mathbb{R}$, the entropy production $\xi \in \mathbb{R}$ and the entropy-flux $\Phi \in \mathbb{R}^3$. Because the specific entropy $s \in \mathbb{R}$ depends on the inner energy $e \in \mathbb{R}$, the deformation gradient $F \in \mathbb{R}^{3\times 3}$, the order parameter $S \in \mathbb{R}$ and its gradient $\nabla S \in \mathbb{R}^3$, we get

$$\xi = \rho_0 \left(\frac{\partial s}{\partial e}\dot{e} + \frac{\partial s}{\partial F} : \dot{F} + \frac{\partial s}{\partial S}\dot{S} + \frac{\partial s}{\partial \nabla S}(\nabla S)^{\cdot} \right) + \operatorname{div}(\Phi).$$

The temperature $\vartheta \in \mathbb{R}$ can be calculated by

$$\vartheta = \frac{\partial e}{\partial s}$$

and thus, we write

$$\xi = \frac{\rho_0}{\vartheta}\dot{e} + \rho_0 \frac{\partial s}{\partial F} : \dot{F} + \rho_0 \frac{\partial s}{\partial S}\dot{S} + \rho_0 \frac{\partial s}{\partial \nabla S}(\nabla S)^{\cdot} + \operatorname{div}(\Phi).$$

Inserting the energy balance

$$\rho_0 \dot{e} + \operatorname{div}(Q) = \dot{F}^{\mathsf{T}} \cdot T$$

with the heat $Q \in \mathbb{R}^3$ and the stress tensor $T \in \mathbb{R}^{3\times 3}$, we get

$$\xi = \frac{1}{\vartheta}\left(-\operatorname{div}(Q) + \dot{F}^{\mathsf{T}} : T\right) + \rho_0 \frac{\partial s}{\partial F} : \dot{F} + \rho_0 \frac{\partial s}{\partial S}\dot{S} + \rho_0 \frac{\partial s}{\partial \nabla S}(\nabla S)^{\cdot} + \operatorname{div}(\Phi)$$

and with

$$-\frac{1}{\vartheta}\operatorname{div}(Q) = -\operatorname{div}\left(\frac{Q}{\vartheta}\right) - \frac{1}{\vartheta^2} Q \cdot \nabla\vartheta$$

this equation reads

$$\xi = -\operatorname{div}\left(\frac{Q}{\vartheta}\right) - \frac{1}{\vartheta^2} Q \cdot \nabla\vartheta + \frac{1}{\vartheta}\dot{F}^{\mathsf{T}} : T + \rho_0 \frac{\partial s}{\partial F} : \dot{F} + \rho_0 \frac{\partial s}{\partial S}\dot{S} + \rho_0 \frac{\partial s}{\partial \nabla S}(\nabla S)^{\cdot} + \operatorname{div}(\Phi).$$

With index notation

$$Q \rightarrow Q_k; \quad T \rightarrow T_{ik}; \quad \dot{F} \rightarrow \dot{F}_{ik}; \quad \nabla \rightarrow \nabla_k; \quad \Phi \rightarrow \Phi_k$$

we write

$$\begin{aligned}\xi &= -\nabla_k \cdot \left(\frac{Q_k}{\vartheta}\right) - \frac{1}{\vartheta^2} Q_k \cdot \nabla_k \vartheta + \frac{1}{\vartheta}\dot{F}_{ki} T_{ik} + \\ &+ \rho_0 \frac{\partial s}{\partial F_{ik}} : \dot{F}_{ik} + \rho_0 \frac{\partial s}{\partial S}\dot{S} + \rho_0 \frac{\partial s}{\partial \nabla_k S} \cdot (\nabla_k S)^{\cdot} + \nabla_k \cdot \Phi_k.\end{aligned}$$

With the product rule for the second last term

$$\nabla_k \left(\frac{\partial}{\partial \nabla_k S}\dot{S}\right) = \nabla_k \left(\frac{\partial}{\partial \nabla_k S}\right)\dot{S} + \left(\frac{\partial}{\partial \nabla_k S}\right)\nabla_k \dot{S}$$

we have

$$\begin{aligned}\xi &= -\nabla_k \cdot \left(\frac{Q_k}{\vartheta}\right) - \frac{1}{\vartheta^2} Q_k \cdot \nabla_k \vartheta + \frac{1}{\vartheta}\dot{F}_{ki} T_{ik} + \\ &+ \rho_0 \frac{\partial s}{\partial F_{ik}} : \dot{F}_{ik} + \rho_0 \frac{\partial s}{\partial S}\dot{S} + \rho_0 \left(\nabla_k \left(\frac{\partial s}{\partial \nabla_k S}\dot{S}\right) - \nabla_k \left(\frac{\partial s}{\partial \nabla_k S}\right)\dot{S}\right) + \nabla_k \cdot \Phi_k.\end{aligned}$$

Sorting the terms yields

$$\begin{aligned}\xi &= \nabla_k \cdot \left(\Phi - \left(\frac{Q_k}{\vartheta}\right) - \frac{\partial s}{\partial \nabla_k S}\dot{S}\right) + \left(\frac{1}{\vartheta} T_{ik} + \rho_0 \frac{\partial s}{\partial F_{ik}}\right)\dot{F}_{ik} + \\ &+ \left(\rho_0 \frac{\partial s}{\partial S} - \rho_0 \nabla_k \left(\frac{\partial s}{\partial \nabla_k S}\right)\right)\dot{S} + Q_k \cdot \frac{\nabla_k}{\vartheta} \geq 0.\end{aligned}$$

The inequality can be fulfilled by the following assumptions

$$\Phi_k = \frac{Q_k}{\vartheta} + \frac{\partial s}{\partial \nabla_k S}\dot{S}, \tag{A.6.1}$$

$$T_{ik} = -\rho_0 \vartheta \frac{\partial s}{\partial F_{ik}}, \quad \vartheta \partial s = \partial e, \tag{A.6.2}$$

$$0 \leq Q_k \cdot \frac{\nabla_k}{\vartheta}, \tag{A.6.3}$$

$$0 \leq \left(\frac{\partial s}{\partial S} - \nabla_k \frac{\partial s}{\partial \nabla_k S}\right)\dot{S}. \tag{A.6.4}$$

- Eq. (A.6.1) denotes the resulting entropy flux.
- Eq. (A.6.2) describes the stresses as a function of the deformations.
- Eq. (A.6.3) is true, since temperature- and heat flux have the same direction.
- Eq. (A.6.4) leads to a phase field equation.

7 Binomial formulas

Lemma 14 (*Formula*)**.** *Let a, b, c be in $R^{n\times n}$, $n \in \mathbb{N}$. Let $a : b = \sum_{ij} a_{ij}b_{ij}$ denote the matrix scalar product. Then we have*

$$\frac{1}{2}(a+b) : (c(a-b)) = a : (c(a-b)) - \frac{1}{2}(a-b) : (c(a-b)) \,. \tag{A.7.1}$$

Proof. We show the lemma by the following calculculation

$$\begin{aligned}
\frac{1}{2}(a+b) : (c(a-b)) &= \frac{1}{2}a : (c(a-b)) + \frac{1}{2}b : (c(a-b)) \\
&= a : (c(a-b)) - \frac{1}{2}a : (c(a-b)) + \frac{1}{2}b : (c(a-b)) \\
&= a : (c(a-b)) - \frac{1}{2}a : (ca) + \frac{1}{2}a : (cb) + \frac{1}{2}b : (ca) - \frac{1}{2}b : (cb) \\
&= a : (c(a-b)) - \frac{1}{2}(a : (ca) - a : (cb) - b : (ca) + b : (cb)) \\
&= a : (c(a-b)) - \frac{1}{2}(a-b) : (c(a-b)) \,.
\end{aligned} \tag{A.7.2}$$

□

If we repeat the proof for a, b, c in R^n, defining $a : b = \sum_i a_i b_i$, omitting the vector c and using the third binomial formula on the left hand side, this yields

$$\frac{1}{2}(a^2 - b^2) = a(a-b) - \frac{1}{2}(a-b)^2 \tag{A.7.3}$$

and with $a =: \nabla S_h^n$ and $b =: \nabla S_h^{n-1}$ we have

$$\frac{1}{2}((\nabla S_h^n)^2 - (\nabla S_h^{n-1})^2) = \nabla S_h^n(\nabla S_h^n - \nabla S_h^{n-1}) - \frac{1}{2}(\nabla S_h^n - \nabla S_h^{n-1})^2 \,. \tag{A.7.4}$$

For $a =: \varepsilon(S_h^{n-1}, u_h^n)$, $b =: \varepsilon(S_h^{n-1}, u_h^{n-1})$ and $c =: \mathbb{C}(S_h^{n-1})$ we achieve with Eq. (A.7.2) and $\varepsilon = \underbrace{\frac{1}{2}(\nabla u_h^{n-1} + (\nabla u_h^{n-1})^\mathsf{T})}_{\varepsilon(u_h^{n-1})} - S_h^{n-1}\bar{\varepsilon}$ the relation

$$\begin{aligned}
\frac{1}{2}(\varepsilon(S_h^{n-1}, u_h^n) &+ \varepsilon(S_h^{n-1}, u_h^{n-1})) : \left(\mathbb{C}(S_h^{n-1})(\varepsilon(S_h^{n-1}, u_h^n) - \varepsilon(S_h^{n-1}, u_h^{n-1}))\right) \\
&= \varepsilon(S_h^{n-1}, u_h^n) : \left(\mathbb{C}(S_h^{n-1})(\varepsilon(S_h^{n-1}, u_h^n) - \varepsilon(S_h^{n-1}, u_h^{n-1}))\right) \\
&- \frac{1}{2}(\varepsilon(S_h^{n-1}, u_h^n) - \varepsilon(S_h^{n-1}, u_h^{n-1})) : \left(\mathbb{C}(S_h^{n-1})(\varepsilon(S_h^{n-1}, u_h^n) - \varepsilon(S_h^{n-1}, u_h^{n-1}))\right) \\
&= \varepsilon(S_h^{n-1}, u_h^n) : \left(\mathbb{C}(S_h^{n-1})(\varepsilon(u_h^n - u_h^{n-1}))\right) \\
&- \frac{1}{2}(\varepsilon(u_h^n - u_h^{n-1}) : \left(\mathbb{C}(S_h^{n-1})(\varepsilon(u_h^n - u_h^{n-1}))\right) \,.
\end{aligned}$$

□

8 Newton's method

Before solving the fully discretised scheme Eq. (9.10) with the Banach fixed-point theorem, we tried the implementation by Newton's method, we will briefly sketch, in case, it will be needed for some future research and implementation.

We want to solve the equation system $F(\hat{S}^n, \hat{u}^{n-1}) = 0$. We calculate

$$\begin{aligned} (\hat{S}^n)^{k+1} &= (\hat{S}^n)^k - \left(dF\left((\hat{S}^n, \hat{u}^{n-1})^k\right)\right)^{-1} F\left((\hat{S}^n, \hat{u}^{n-1})^k\right), \\ (\hat{S}^n)^0 &= \hat{S}^n . \end{aligned}$$

Beginning with the actual time step, the next time step will start after the Newton-Iteration fulfills an error criterion

$$||(\hat{S}^n)^{k+1} - (\hat{S}^n)^k|| \leq ||e|| \tag{A.8.1}$$

with a given error e. The terms of F are calculated per element by

$$F = \sum_{el} F^{(el)} \quad \in \mathbb{R}^N; \quad \left(F^{(el)}_{loc}\right)_i = \begin{pmatrix} F_1^{(el)} \\ F_2^{(el)} \\ F_3^{(el)} \\ F_4^{(el)} \end{pmatrix} \quad \in \mathbb{R}^4$$

with

$$\begin{aligned} F_j^{(el)}\left((S^{(el)}_{loc,h})^n\right) &= \int_{\Omega_{el}} c\left(|\sum_i B_i \hat{S}_i^{n-1}|\right) N_j \left(\sum_i N_i \hat{S}_i^n - \sum_i N_i \hat{S}_i^{n-1}\right) \\ &\quad + \tau\, N_j W_S\left(\text{arg}_1\right) + \tau\beta\, N_j \psi'\left(\text{arg}_2\right) + \tau\alpha \sum_i K_{ij} \hat{S}_j^n \, dx_{el} , \end{aligned}$$

$\text{arg}_1 = \sum N_i \hat{S}_i^n; \sum N_i \hat{S}_i^{n-1}, u_h^{n-1}$ and $\text{arg}_2 = \sum_i N_i \hat{S}_i^n; \sum_i N_i \hat{S}_i^{n-1}$.

The matrix containing the derivatives of F_I with respect to $\hat{S}_J$ is called Jacobian matrix and it is calculated (element-wise) by

$$\left(dF(\hat{S})\right)_{IJ} = \frac{\partial F_J(\hat{S})}{\partial S_I} = \sum_{el} dF^{(el)}(\hat{S}) = \sum_{el} \frac{\partial F_i}{\partial S_j}(\hat{S})$$

with

$$\left(dF^{(el)}_{loc}\right)_{ji} = \begin{pmatrix} dF^{(el)}_{11} & dF^{(el)}_{12} & dF^{(el)}_{13} & dF^{(el)}_{14} \\ dF^{(el)}_{21} & dF^{(el)}_{22} & dF^{(el)}_{23} & dF^{(el)}_{24} \\ dF^{(el)}_{31} & dF^{(el)}_{32} & dF^{(el)}_{33} & dF^{(el)}_{34} \\ dF^{(el)}_{41} & dF^{(el)}_{42} & dF^{(el)}_{34} & dF^{(el)}_{44} \end{pmatrix} \in \mathbb{R}^{4\times 4}.$$

The components are calculated by

$$\begin{aligned} dF^{(el)}_{ji}\left((S^{(el)}_{loc,h})^{n-1}\right) &= \int_{\Omega_{el}} \left(\left(|\sum_i B_i \hat{S}^{n-1}_i|\right) N_i N_j + \tau\, W_{SS}(\arg_1) N_i N_j \right. \\ &\quad \left. + \tau\beta\, \psi^{''}(\arg_2) N_i N_j + \tau\alpha \sum_i K_{ij}\, dx_{el} \qquad \text{with} \right. \end{aligned}$$

$$\begin{aligned} W_{js} &= e_{j1}\,(N_i\hat{S}^n_i)^2 + e_{j2}\,N_i\hat{S}^n_i + e_{j3}\,, \\ W_{jss} &= 2e_{j1}\,(N_i\hat{S}^n_i) + e_{j2}\,, \\ \psi^{'}_j &= (N_i\hat{S}^n_i)^3 + d_{j1}\,(N_i\hat{S}^n_i)^2 + d_{j2}\,N_i\hat{S}^n_i + d_{j3}\,, \\ \psi^{''}_j &= 3((N_i\hat{S}^n_i)^2 + d_{j1}\,2N_i\hat{S}^n_i + d_{j2}\,, \\ d_{j1} &= -2 + N_i\hat{S}^{n-1}_i\,, \\ d_{j2} &= 1 - 2N_i\hat{S}^{n-1}_i + (N_i\hat{S}^{n-1}_i)^2\,, \\ d_{j3} &= N_i\hat{S}^{n-1}_i - 2(N_i\hat{S}^{n-1}_i)^2 + (N_i\hat{S}^{n-1}_i)^3\,, \\ e_{j1} &= \frac{1}{2}\,\bar{\varepsilon}((\mathbb{C}_2 - \mathbb{C}_1)\bar{\varepsilon})\,, \\ e_{j2} &= -\bar{\varepsilon}((\mathbb{C}_2 - \mathbb{C}_1)(\varepsilon(u^{n-1}_h)_j) + \frac{1}{2}\,\bar{\varepsilon}(\mathbb{C}_1\bar{\varepsilon}) + \frac{1}{2}\,N_i\hat{S}^{n-1}_i\bar{\varepsilon}((\mathbb{C}_2 - \mathbb{C}_1)\bar{\varepsilon})\,, \\ e_{j3} &= (\frac{1}{2}\,\varepsilon(u^{n-1}_h)_j - \bar{\varepsilon}N_i\hat{S}^{n-1}_i)((\mathbb{C}_2 - \mathbb{C}_1)\varepsilon(u^{n-1}_h)_j) - \bar{\varepsilon}(\mathbb{C}_1\varepsilon(u^{n-1}_h)_j) \\ &\quad + \frac{1}{2}\,N_i\hat{S}^{n-1}_i\bar{\varepsilon}(\mathbb{C}_1\bar{\varepsilon}) + \frac{1}{2}\,\bar{\varepsilon}((\mathbb{C}_2 - \mathbb{C}_1)\bar{\varepsilon}\,(N_i\hat{S}^{n-1}_i)^2)\,. \end{aligned}$$

$$K_{ij} = B_i B_j\,; \quad B_i = \begin{pmatrix} N_{i,x} \\ N_{i,y} \end{pmatrix}; \quad \varepsilon(u_h)_j = \frac{1}{2}\left((\nabla u_h)_j + (\nabla u_h)^T_j\right), \quad \nabla(u_h)_j = B^u_j \hat{u}_j\,.$$

9 Piezoelectricity

The structure of a piezoelectric system is almost identical to the equation system of linear elasticity. To transfer a phase field model to piezoelectricity the terms shown in table A.9.2 have to be replaced.

Table A.9.2: Analogies of linear elasticity and piezoelectricity.

1	Description	Elasticity	Piezo
2	Balance law	$\mathrm{div}\sigma = f$	$\mathrm{div}\sigma = f$
3	Balance law	-	$\mathrm{div}D = \rho$
4	Stress tensor	$\sigma = \partial_\varepsilon W$	$\sigma = \partial_\varepsilon W$
5	Stress tensor	$\sigma = \mathbb{C}\varepsilon$	$\sigma = \mathbb{C}\varepsilon - \mathbb{E}E$
6	Strain tensor	$\varepsilon = \frac{1}{2}\left(\nabla u + (u)^{\mathsf{T}}\right)$	$\varepsilon = \mathbb{C}^{-1}\sigma + \mathbb{D}E$
7	Electric field	-	$E = \nabla\varphi$
8	Electric displacement	-	$D = -\partial_E H;\ D = \xi E + \mathbb{E}\varepsilon + P^0$
9	Enthalpy / mech. work	$W = \frac{1}{2}\varepsilon^{\mathsf{T}}\mathbb{C}\varepsilon$	$H = \frac{1}{2}\varepsilon^{\mathsf{T}}\mathbb{C}\varepsilon - \varepsilon(\mathbb{E}^{\mathsf{T}}E) - \frac{1}{2}E(\xi E) - P^0 E$
10	Polarisation	order parameter	$P = \mathbb{D}\sigma + \mathbb{K}E$

Table A.9.3: Notation.

$\mathbb{C}$:	Elasticity tensor	$\mathbb{K}$:	Tensor of susceptibility
D :	Electric displacement	ξ :	Electric field constant
$\mathbb{D}$:	Piezoelectric tensor	P^0 :	Spontaneous polarisation
E :	Electric field	ρ :	External electric load
$\mathbb{E}$:	Piezoelectric tensor	σ :	Strain tensor
ε :	Mechanical displacement	φ :	Electric potential
f :	External force	u :	Displacement
H :	Electric enthalpy	W :	Mechanical work

BIBLIOGRAPHY

[1] Hans-Dieter Alber. "A model for brittle fracture based on the hybrid phase field model". In: *Contin. Mech. Thermodyn.* 24.4-6 (2012), pp. 391–402. ISSN: 0935-1175. DOI: 10.1007/s00161-011-0211-z. URL: https://doi.org/10.1007/s00161-011-0211-z.

[2] Hans-Dieter Alber. "An alternative to the Allen-Cahn phase field model for interfaces in solids—numerical efficiency". In: *Continuous media with microstructure 2*. Springer, 2016, pp. 121–136.

[3] Hans-Dieter Alber. "Asymptotics and numerical efficiency of the Allen-Cahn model for phase interfaces with low energy in solids". In: *Contin. Mech. Thermodyn.* 29.3 (2017), pp. 757–803. ISSN: 0935-1175. DOI: 10.1007/s00161-017-0558-x. URL: https://doi.org/10.1007/s00161-017-0558-x.

[4] Hans-Dieter Alber and Peicheng Zhu. "Comparison of a rapidely converging phase field model for interfaces in solids with the Allen-Cahn model". In: *J. Elasticity* 111.2 (2013), pp. 153–221. ISSN: 0374-3535. DOI: 10.1007/s10659-012-9398-x. URL: https://doi.org/10.1007/s10659-012-9398-x.

[5] Hans-Dieter Alber and Peicheng Zhu. "Solutions to a model with Neumann boundary conditions for phase transitions driven by configurational forces". In: *Nonlinear Analysis-real World Applications - NONLINEAR ANAL-REAL WORLD APP* 12 (2011). DOI: 10.1016/j.nonrwa.2010.11.012.

[6] Hans-Dieter Alber and Peicheng Zhu. "Solutions to a model with nonuniformly parabolic terms for phase evolution driven by configurational forces". In: *SIAM J. Appl. Math.* 66.2 (2005), pp. 680–699. ISSN: 0036-1399. DOI: 10.1137/050629951. URL: https://doi.org/10.1137/050629951.

[7] Samuel M. Allen and John W. Cahn. "A microscopic theory for antiphase boundary motion and its application to antiphase domain coarsening". In: *Acta Metallurgica* 27 (1979), pp. 1085–1095.

[8] Samuel M. Allen and John W. Cahn. "Coherent and incoherent equilibria in iron-rich iron-aluminium alloys". English. In: *Acta Metallurgica* 23.9 (1975), pp. 1017–1026. ISSN: 0001-6160. DOI: 10.1016/0001-6160(75)90106-6.

[9] Robert F. Almgrem. "Second-order phase field asymptotics for unequal conductivities". In: *SIAM J. Appl. Math* 59.6 (1999), pp. 2086–2107.

[10] Vittorio E. Badalassi, Hector D. Ceniceros, and Sanjoy Banerjee. "Computation of multiphase systems with phase field models". In: *J. Comput. Phys.* 190.2 (2003), pp. 371–397. ISSN: 0021-9991. DOI: 10.1016/S0021-9991(03)00280-8. URL: https://doi.org/10.1016/S0021-9991(03)00280-8.

[11] Hans-Jürgen Bargel and Günter Schulze. *Werkstoffkunde*. Berlin Heidelberg New York: Springer-Verlag, 2013. ISBN: 978-3-642-17717-0.

[12] John W. Barret, Harald Garcke, and Robert Nürnberg. "Finite element approximation of a phase field model for void electromigration". In: *SIAM J. Numer. Anal* (2004), p. 772.

[13] John W. Barrett and James F. Blowey. "Finite element approximation of the Cahn-Hilliard equation with concentration dependent mobility". In: *Math. Comp.* 68.226 (1999), pp. 487–517. ISSN: 0025-5718. DOI: 10.1090/S0025-5718-99-01015-7. URL: https://doi.org/10.1090/S0025-5718-99-01015-7.

[14] Sören Bartels. *Numerical Methods for Nonlinear Partial Differential Equations*. Berlin, Heidelberg: Springer-Verlag, 2015. ISBN: 978-3-319-13797-1.

[15] Klaus-Jürgen Bathe. *Finite-Elemente-Methoden*. Berlin, Heidelberg: Springer-Verlag, 2008. ISBN: 978-3-540-66806-0.

[16] Kaushik Bhattacharya and Guruswaminaidu Ravichandran. "Ferroelectric perovskites for electromechanical actuation". In: *Acta Materialia* 51 (2003), pp. 5941–5960.

[17] Luise Blank, Harald Garcke, Lavinia Sarbu, Tarin Srisupattarawanit, Vanessa Styles, and Axel Voigt. "Phase-field approaches to structural topology optimization". In: *Constrained optimization and optimal control for partial differential equations*. Vol. 160. Internat. Ser. Numer. Math. Birkhäuser/Springer Basel AG, Basel, 2012, pp. 245–256. DOI: 10.1007/978-3-0348-0133-1_13. URL: https://doi.org/10.1007/978-3-0348-0133-1_13.

[18] Luise Blank, Harald Garcke, Lavinia Sarbu, and Vanessa Styles. "Primal-dual active set methods for Allen-Cahn variational inequalities with nonlocal constraints". In: *Numerical Methods for Partial Differential Equations* 29.3 (2013), pp. 999–1030. DOI: 10.1002/num.21742. eprint: https://onlinelib-rary.wiley.com/doi/pdf/10.1002/num.21742.

URL: https://onlinelibrary.wiley.com/doi/abs/10.1002/num.21742.

[19] Thomas Blesgen and Ulrich Weikard. "Multi-component Allen-Cahn equation for elastically stressed solids". In: *Electron. J. Differential Equations* 89 (2005), pp. 1–17. ISSN: 1072-6691.

[20] James F. Blowey and Charles M. Elliott. "Curvature dependent phase boundary motion and parabolic double obstacle problems". In: *Degenerate diffusions*. Vol. 47. IMA Vol. Math. Appl. Springer-Verlag, New York, 1993, pp. 19–60. DOI: 10.1007/978-1-4612-0885-3_2. URL: https://doi.org/10.1007/978-1-4612-0885-3_2.

[21] Elena Bonetti, Pierluigi Colli, Wolfgang Dreyer, Gianni Gilardi, Giulio Schimperna, and Jürgen Sprekels. "On a model for phase separation in binary alloys driven by mechanical effects". In: *Physica D. Nonlinear Phenomena* 165.1-2 (2002), pp. 48–65. ISSN: 0167-2789. DOI: 10.1016/S0167-2789(02)00373-1. URL: https://doi.org/10.1016/S0167-2789(02)00373-1.

[22] Anke Böttcher and Herbert Egger. "Structure Preserving Discretization of Allen-Cahn Type Problems Modeling the Motion of Phase Boundaries". In: *Vietnam Journal of Mathematics* (2020), pp. 2305–2228. DOI: 10.1007/s10013-020-00428-w. URL: https://doi.org/10.1007/s10013-020-00428-w.

[23] Dietrich Braess. *Finite Elemente - Theorie, schnelle Löser und Anwendungen in der Elastizitätstheorie*. Berlin Heidelberg New York: Springer-Verlag, 2007. ISBN: 978-3-540-72449-0.

[24] Gunduz Caginalp and Xinfu Chen. "Convergence of the phase field model to its sharp interface limits". In: *European J. Appl. Math.* 9.4 (1998), pp. 417–445. ISSN: 0956-7925. DOI: 10.1017/S0956792598003520. URL: https://doi.org/10.1017/S0956792598003520.

[25] Wenwu Cao. "Phenomenological theories of ferroelectric phase transitions". In: *British Ceramic Transactions* 103.2 (2004), pp. 71–75.

[26] Yun Gang Chen, Yoshikazu Giga, and Shun'ichi Goto. "Uniqueness and existence of viscosity solutions of generalized mean curvature flow equations". In: *J. Differential Geom.* 33.3 (1991), pp. 749–786. ISSN: 0022-040X. URL: http://projecteuclid.org/euclid.jdg/1214446564.

[27] Yongho Choi, Darae Jeong, Seunggyu Lee, Minhyun Yoo, and Junseok Kim. "Motion by mean curvature of curves on surfaces using the Allen-Cahn equation". In: *Internat. J. Engrg. Sci.* 97 (2015), pp. 126–132. ISSN: 0020-7225. DOI: 10.1016/j.ijengsci.2015.10.002. URL: https://doi.org/10.1016/j.ijengsci.2015.10.002.

[28] Philippe G. Ciarlet. *Mathematical Elasticity. Volume I: Three-Dimensional Elasticity*. Vol. 20. Elsevier, 1988. ISBN: 9780444817761.

[29] Tobias Holck Colding, William P. Minicozzi, and Erik Kjær Pedersen. "Mean curvature flow". In: *Bull. Amer. Math. Soc. (N.S.)* 52.2 (2015), pp. 297–333. ISSN: 0273-0979. DOI: 10.1090/S0273-0979-2015-01468-0. URL: https://doi.org/10.1090/S0273-0979-2015-01468-0.

[30] Patricia Derksen. *Die 2. Kornsche Ungleichung*. Diplomarbeit Fakultät für Mathematik, Universität Duisburg Essen, 2013.

[31] Manfred Dobrowolski. *Angewandte Funktionalanalysis*. 2nd ed. Springer-Verlag Berlin Heidelberg, 2010. ISBN: 978-3-642-15269-6. DOI: 10.1007/978-3-642-15269-6.

[32] Marc Droske and Martin Rumpf. "A level set formulation for Willmore flow". In: *Interfaces Free Bound.* 6.3 (2004), pp. 361–378. ISSN: 1463-9963. DOI: 10.4171/IFB/105. URL: https://doi.org/10.4171/IFB/105.

[33] Qiang Du and Roy A. Nicolaides. "Numerical analysis of a continuum model of phase transition". In: *SIAM J. Numer. Anal.* 28.5 (1991), pp. 1310–1322. ISSN: 0036-1429. DOI: 10.1137/0728069. URL: https://doi.org/10.1137/0728069.

[34] Christof Eck, Harald Garcke, and Peter Knabner. *Mathematische Modellierung*. Berlin Heidelberg New York: Springer-Verlag, 2011. ISBN: 978-3-662-54335-1.

[35] Lawrence C. Evans. *Partial Differential Equations*. Heidelberg: American Mathematical Soc., 2010. ISBN: 978-0-821-84974-3.

[36] Lawrence C. Evans, H. Mete Soner, and Panagiotis E. Souganidis. "Phase transitions and generalized motion by mean curvature". In: *Comm. Pure Appl. Math.* 45.9 (1992), pp. 1097–1123. ISSN: 0010-3640. DOI: 10.1002/cpa.3160450903. URL: https://doi.org/10.1002/cpa.3160450903.

[37] Lawrence C. Evans and Joel Spruck. "Motion of level sets by mean curvature. I". In: *J. Differential Geom.* 33.3 (1991), pp. 635–681. ISSN: 0022-040X. URL: http://projecteuclid.org/euclid.jdg/1214446559.

[38] Xiaobing Feng and Andreas Prohl. "Numerical analysis of the Allen-Cahn equation and approximation for mean curvature flows". In: *Numer. Math.* 94.1 (2003), pp. 33–65. ISSN: 0029-599X. DOI: 10.1007/s00211-002-0413-1. URL: https://doi.org/10.1007/s00211-002-0413-1.

[39] Xiaobing Feng and Hai-Jun Wu. "A posteriori error estimates and an adaptive finite element method for the Allen-Cahn equation and the mean curvature flow". In: *J. Sci. Comput.* 24.2 (2005), pp. 121–146. ISSN: 0885-7474. DOI: 10.1007/s10915-004-4610-1. URL: https://doi.org/10.1007/s10915-004-4610-1.

[40] Xinlong Feng, Huailing Song, Tao Tang, and Jiang Yang. "Nonlinear stability of the implicit-explicit methods for the Allen-Cahn equation". In: *Inverse Probl. Imaging* 7.3 (2013), pp. 679–695. ISSN: 1930-8337. DOI: 10.3934/ipi.2013.7.679. URL: https://doi.org/10.3934/ipi.2013.7.679.

[41] Xinlong Feng, Tao Tang, and Jiang Yang. "Stabilized Crank-Nicolson/Adams-Bashforth schemes for phase field models". In: *East Asian J. Appl. Math.* 3.1 (2013), pp. 59–80. ISSN: 2079-7362. DOI: 10.4208/eajam.200113.220213a. URL: https://doi.org/10.4208/eajam.200113.220213a.

[42] Peter Fratzl, Oliver Penrose, and Joel L. Lebowitz. "Modeling of phase separation in alloys with coherent elastic misfit". In: *J. Statist. Phys.* 95.5-6 (1999), pp. 1429–1503. ISSN: 0022-4715. DOI: 10.1023/A:1004587425006. URL: https://doi.org/10.1023/A:1004587425006.

[43] Eliot Fried and Morton E. Gurtin. "Dynamic solid-solid transitions with phase characterized by an order parameter". In: *Phys. D* 72.4 (1994), pp. 287–308. ISSN: 0167-2789. DOI: 10.1016/0167-2789(94)90234-8. URL: https://doi.org/10.1016/0167-2789(94)90234-8.

[44] Harald Garcke. "Curvature driven interface evolution". In: *Jahresber. Dtsch. Math.-Ver.* 115.2 (2013), pp. 63–100. ISSN: 0012-0456. DOI: 10.1365/s13291-013-0066-2. URL: https://doi.org/10.1365/s13291-013-0066-2.

[45] Harald Garcke. "On a Cahn-Hilliard model for phase separation with elastic misfit". In: *Annales de l'Institut Henri Poincare (C) Non Linear Analysis* 22 (2005), pp. 165–185. DOI: 10.1016/j.anihpc.2004.07.001.

[46] Harald Garcke. "On Cahn-Hilliard Systems with Elasticity". In: *Royal Society of Edinburgh - Proceedings A* 133A (2003), pp. 307–331. DOI: 10.1017/S03082105-00002419.

[47] Harald Garcke and Björn Stinner. "Second order phase field asymptotics for multi-component systems". In: *Interfaces Free Bound.* 8.2 (2006), pp. 131–157. ISSN: 1463-9963. DOI: 10.4171/IFB/138. URL: https://doi.org/10.4171/IFB/138.

[48] Vitaly L. Ginzburg. "On superconductivity and superfluidity (what I have and have not managed to do), as well as on the 'physical minimum' at the beginning of the XXI century (December 8, 2003)". In: *Physics-Uspekhi* 47.11 (2004), pp. 1155–1170. ISSN: 1063-7869. DOI: 10.1070/PU2004v047n11ABEH001825.

[49] Carsten Gräser. *Analysis und Approximation der Cahn-Hilliard Gleichung mit Hindernispotential.* Diplomarbeit FB Mathematik und Informatik, FU Berlin, 2004.

[50] Dietmar Gross, Werner Hauger, and Peter Wriggers. *Technische Mechanik 4 - Hydromechanik, Elemente der Höheren Mechanik, Numerische Methoden.* Berlin Heidelberg New York: Springer-Verlag, 1993. ISBN: 978-3-642-41000-0.

[51] Lars Grüne and Oliver Junge. *Gewöhnliche Differentialgleichungen.* Springer Spektrum, 2016. ISBN: 2364-2378. DOI: 10.1007/978-3-658-10241-8.

[52] Stefan Hartmann. *Finite-Elemente Berechnung inelastischer Kontinua.* Habilitationsschrift, Fachgebiet Mechanik, Universität Kassel, 2003.

[53] Peter Haupt. *Continuum Mechanics and Theory of Materials.* Berlin Heidelberg New York: Springer-Verlag, 2000. ISBN: 3-540-66114-X.

[54] Claudia Hecht. *Existence theory and necessary optimality conditions for the control of the elastic Allen-Cahn system.* Diplomarbeit Universität Regensburg, 2011.

[55] Ulrich Heisserer. "High-order finite elements for material and geometric nonlinear finite strain problems". PhD thesis. 2008. ISBN: 978-3-8322-7703-1.

[56] Thomas Y. Hou, Phoebus Rosakis, and Philippe LeFloch. "A level-set approach to the computation of twinning and phase-transition dynamics". In: *J. Comput. Phys.* 150.2 (1999), pp. 302–331. ISSN: 0021-9991. DOI: 10.1006/jcph.1998.6179. URL: https://doi.org/10.1006/jcph.1998.6179.

[57] Gerhard Huisken. "Asymptotic behavior for singularities of the mean curvature flow". In: *J. Differential Geom.* 31.1 (1990), pp. 285–299. ISSN: 0022-040X. URL: http://projecteuclid.org/euclid.jdg/1214444099.

[58] Darae Jeong, Seunggyu Lee, Dongsun Lee, Jaemin Shin, and Junseok Kim. "Comparison study of numerical methods for solving the Allen-Cahn equation". In: *Comput. Mater. Science* 111 (2016), pp. 131–136.

[59] Alain Karma and Wouter-Jan Rappel. "Quantitative phase-field modeling of dendritic growth in two and three dimensions". In: *Physical Review E* 57.4 (1998), pp. 4323–4349.

[60] Daniel Kessler, Ricardo H. Nochetto, and Alfred Schmidt. "A posteriori error control for the Allen-Cahn problem: circumventing Gronwall's inequality". In: *M2AN Math. Model. Numer. Anal.* 38.1 (2004), pp. 129–142. ISSN: 0764-583X. DOI: 10.1051/m2an:2004006. URL: https://doi.org/10.1051/m2an:2004006.

[61] Dilip Kondepudi and Ilya Prigogine. *Modern Thermodynamics - From Heat Engines to Dissipative Structures*. New York: John Wiley and Sons, 2014. ISBN: 978-1-118-69870-9.

[62] Charlotte Kuhn and Ralf Müller. "A continuum phase field model for fracture". In: *Eng. Fract. Mech.* 77.18, SI (2010), pp. 3625–3634.

[63] Lev D. Landau. "Theory of phase transformations. II". In: *Phys. Z. Sowjetunion 11* 545 (1937).

[64] Lev D. Landau. "Theory of supraconductivity". In: *Phys. Z. Sowjetunion 11* 129 (1937).

[65] Lev Davidovich Landau. "Theory of phase transformations. I". In: *Phys. Z. Sowjetunion 11* 26 (1937).

[66] Dong S. Lee and Junseo S. Kim. "Mean curvature flow by the Allen-Cahn equation". In: *European J. Appl. Math.* 26.4 (2015), pp. 535–559. ISSN: 0956-7925. DOI: 10.1017/S0956792515000200. URL: https://doi.org/10.1017/S0956792515000200.

[67] Perry H. Leo, John S. Lowengrub, and Herng Jeng Jou. "A diffuse interface model for microstructural evolution in elastically stressed solids". In: *Acta Materialia* 46.6 (1998), pp. 2113–2130.

[68] Andreas Liehr. *Dissipative Solitons in Reaction Diffusion Systems - Mechanisms, Dynamics, Interaction*. Berlin Heidelberg: Springer Science and Business Media, 2013. ISBN: 978-3-642-31251-9.

[69] Santiago Madruga, Jie Shen, and Xiaofeng Yang. "Numerical approximations of Allen-Cahn and Cahn-Hilliard equations". In: *Discrete Contin. Dyn. Syst.* 28.4 (2010), pp. 1669–1691. ISSN: 1078-0947. DOI: 10.3934/dcds.2010.28.1669. URL: https://doi.org/10.3934/dcds.2010.28.1669.

[70] Nele Moelans, Bart Blanpain, and Patrick Wollants. "An introduction to phase-field modeling of microstructure evolution". In: *Calphad-Computer Coupling of Phase Diagrams and Thermochemistry* 32.2 (2008), pp. 268–294.

[71] Ralf Müller. *Configurational forces in defect mechanics and in compuational methods*. Habilitation thesis, Fachbereich Mechanik, TU Darmstadt, 2005.

[72] Ralf Müller. "Drei-D-Simulation der Mikrostrukturentwicklung in Zwei-Phasen-Materialien". PhD thesis. Darmstadt: TU Darmstadt, Jan. 2001. URL: http://tubiblio.ulb.tu-darmstadt.de/15675/.

[73] Ralf Müller, Anke Böttcher, Baixiang Xu, Jan Aurich, and Dietmar Gross. "Driving forces on interfaces in elastic-plastic two phase materials". In: *ZAMM Z. Angew. Math. Mech.* 90.10-11, Special Issue: On the Occasion of Zenon Mróz' 80th Birthday (2010), pp. 812–820. ISSN: 0044-2267. DOI: 10.1002/zamm.201000051. URL: https://doi.org/10.1002/zamm.201000051.

[74] Ralf Müller and Dietmar Gross. "Modelling of Microstructure Evolution in Two-Phase Materials using Generalized Driving Forces". In: *Pamm* 1 (Mar. 2002). DOI: 10.1002/1617-7061(200203)1:1<20::AID-PAMM20>3.0.CO;2-J.

[75] Andrew M. Mullis, Christopher E. Goodyer, and Peter K. Jimack. "Towards a 3-Dimensional Phase-Field Model of Dendritic Solidification width Physically Realistic Interface Width". In: *Trans. Indian. Inst. Met.* 65.6 (2012).

[76] Jacob Rubinstein, Peter Sternberg, and Joseph B. Keller. "Fast reaction, slow diffusion, and curve shortening". In: *SIAM J. Appl. Math.* 49.1 (1989), pp. 116–133. ISSN: 0036-1399. DOI: 10.1137/0149007. URL: https://doi.org/10.1137/0149007.

[77] Regina Schmitt, Charlotte Kuhn, and Ralf Müller. "On a phase field approach for martensitic transformations in a crystal plastic material at a loaded surface". In: *Contin. Mech. Thermodyn.* 29.4 (2017), pp. 957–968. ISSN: 0935-1175. DOI: 10.1007/s00161-015-0446-1. URL: https://doi.org/10.1007/s00161-015-0446-1.

[78] Regina Schmitt, Ralf Müller, Charlotte Kuhn, and Herbert M. Urbassek. "A phase field approach for multivariant martensitic transformations of stable and metastable phases". In: *Arch. of Appl. Mech.* 83.6 (2013), pp. 849–859.

[79] David Schrade, Ralf Müller, Dietmar Gross, Marc-Andre Keip, Huy N.M. Thai, and J. Schröder. "An invariant formulation for phase field models in ferroelectrics". In: *Int. J. Solids Structures* 51.11-12 (2014), pp. 2144–2156.

[80] David Schrade, Ralf Müller, Bai-Xiang Xu, and Dietmar Gross. "Domain evolution in ferroelectric materials: A continuum phase field model and finite element implementation". In: *Comp. Meth. Appl. Mech. Eng.* 196.41-44 (2007), pp. 4365–4374.

[81] Ben Schweizer. *Partielle Differentialgleichungen: Eine anwendungsorientierte Einführung*. Springer-Lehrbuch Masterclass. Springer Berlin Heidelberg, 2013. ISBN: 9783642406379. URL: https://books.google.de/books?id=X58InwEACAAJ.

[82] Jaemin Shin, Seong-Kwan Park, and Junseok Kim. "A hybrid FEM for solving the Allen-Cahn equation". In: *Appl. Math. Comput.* 244 (2014), pp. 606–612. ISSN: 0096-3003. DOI: 10.1016/j.amc.2014.07.040. URL: https://doi.org/10.1016/j.amc.2014.07.040.

[83] Tomas Sluka, Kyle G. Webber, Enrico Colla, and Dragan Damjanovic. "Phase field simulations of ferroelastic toughening: The influence of phase boundaries and domain structures". In: *Acta Materialia* 60.13-14 (2012), pp. 5172–5181.

[84] Halil Mete Soner. "Ginzburg-Landau equation and motion by mean curvature. I. Convergence". In: *J. Geom. Anal.* 7.3 (1997), pp. 437–475. ISSN: 1050-6926. DOI: 10.1007/BF02921628. URL: https://doi.org/10.1007/BF02921628.

[85] Yu Su and Chad M. Landis. "Continuum thermodynamics of ferroelectric domain evolution: theory, finite element implementation, and application to domain wall pinning". In: *J. Mech. Phys. Solids* 55.2 (2007), pp. 280–305. ISSN: 0022-5096. DOI: 10.1016/j.jmps.2006.07.006. URL: https://doi.org/10.1016/j.jmps.2006.07.006.

[86] Tao Tang and Jiang Yang. "Implicit-explicit scheme for the Allen-Cahn equation preserves the maximum principle". In: *J. Comput. Math.* 34.5 (2016), pp. 471–481. ISSN: 0254-9409. DOI: 10.4208/jcm.1603-m2014-0017. URL: https://doi.org/10.4208/jcm.1603-m2014-0017.

[87] Yangxin Tang, Wenhua Wang, and Yu Zhou. "A new proof of the existence of weak solutions to a model for phase evolution driven by material forces". In: *Mathematical Methods in the Applied Sciences* 40.13 (2017), pp. 4880–4891. DOI: 10.1002/mma.4354. eprint: https://onlinelibrary.wiley.com/doi/pdf/10.1002/mma.4354. URL: https://onlinelibrary.wiley.com/doi/abs/10.1002/mma.4354.

[88] Robert L. Taylor. *FEAP – A Finite Element Analysis Program*. University of California at Berkeley, Example Manual Version 8.4, 2013.

[89] Bai-Xiang Xu, David Schrade, Dietmar Gross, and Ralf Müller. "Phase field simulation of domain structures in cracked ferroelectrics". In: *Int. J. Fracture* 165.2 (2010), pp. 163–173.

[90] Jian Zhang and Qiang Du. "Numerical studies of discrete approximations to the Allen-Cahn equation in the sharp interface limit". In: *SIAM J. Sci. Comput.* 31.4 (2009), pp. 3042–3063. ISSN: 1064-8275. DOI: 10.1137/080738398. URL: https://doi.org/10.1137/080738398.

[91] Peicheng Zhu. "Regularity of solutions to a model for solid-solid phase transitions driven by configurational forces". In: *Journal of Mathematical Analysis and Applications* 389.2 (2012), pp. 1159–1172. ISSN: 0022-247X. DOI: https://doi.org/10.1016/j.jmaa.2011.12.052. URL: http://www.sciencedirect.com/science/article/pii/S0022247X11011887.

[92] Peicheng Zhu. "Solvability via viscosity solutions for a model of phase transitions driven by configurational forces". In: *Journal of Differential Equations* 251.10 (2011), pp. 2833–2852. ISSN: 0022-0396. DOI: https://doi.org/10.1016/j.jde.2011.05.035. URL: http://www.sciencedirect.com/science/article/pii/S0022039611002178.

Curriculum Vitae

Personal details	
name	Anke Böttcher
date of birth	18.07.1967
place of birth	Berlin-Tempelhof
marital status	married / 4 children
nationality	german
Education	
1973 - 1977	Grundschule Münster/NRW
1977 - 1986	Schillergymnasium Münster/NRW (Abitur)
1983 - 1984	Highschool Galena, Illinois, USA (highschool diploma)
Professional education	
1988 - 1990	photographer (examination), Emsdetten
1992 - 1994	motor mechanic (examination), Kassel
Academic studies	
1986 - 1988	romance philology, University Münster/NRW
1995 - 2003	mechanical engineering (diploma I), University of Kassel
2003 - 2006	mechanical engineering (diploma II), University of Kassel
Professional career	
1990 - 1992	freelance photographer, Münster/NRW
1994 - 1995	employed metal worker, Kassel
1997 - 2006	student assistant, University of Kassel
2006 - 2014	research assistant, TU Darmstadt
2009 - 2010	research assistant, University of Kaiserslautern
2014 - 2016	coordinator profile area, TU Darmstadt
2016 - 2017	phd student, TU Darmstadt
2017 - today	employee of medico international e.V., Frankfurt